Better Things

Published in Great Britain in 2023 by

Laurence King Student & Professional
An imprint of Quercus Editions Ltd
Carmelite House
50 Victoria Embankment
London EC4Y 0DZ

An Hachette UK company

A CIP catalogue record for this book is available from the British Library

TPB ISBN 978-1-52941-968-9
EBOOK ISBN 978-1-52941-969-6

10 9 8 7 6 5 4 3 2 1

Commissioning editor: Liz Faber
Project editor: Angela Koo
Design concept: Daniel Liden
Layout design: Jane Lanaway

Printed and bound in Malaysia

For my daughter Lärke, for daring to ask big questions like 'What's everything made of?'

And for George, who always dared an attempt at answering big questions.
I miss you, my friend.

Better Things

Materials for sustainable product design

Daniel Liden

Laurence King Publishing

Contents

Foreword 6
Introduction 8

CHAPTER 1: PLASTICS **16**
Plastic recycling **20**
Interview: Efrat Friedland *24*
Recycled polypropylene 28
Recycled polyethylene 30
Recycled polyethylene terephthalate 32
Recycled acrylonitrile butadiene styrene 34
Recycled polycarbonate 36
Recycled polyamide 38
Recycled thermoplastic elastomers 40
Recycled silicone rubber 42
Recycled natural rubber 44
Renewable plastics **46**
Interview: Pilar Bolumburu *50*
Renewable polypropylene 54
Renewable polyethylene 56
Polylactide 58
Cellulose acetate 60
Renewable polycarbonate 62
Renewable polyamide 64
Renewable liquid epoxy resin 66
Renewable thermoplastic elastomers 68
Natural rubber 70

CHAPTER 2: TEXTILES **72**
Textile recycling **74**
Interview: Michael Wolf *76*
Recycled polyester textiles 80
Recycled polyamide textiles 82
Recycled cellulose-based synthetic textiles 84
Recycled cotton textiles 86
Recycled wool textiles 88
Recycled leather 90
Renewable textiles **92**
Interview: Charlotte McCurdy *94*
Renewable polyester textiles 98
Renewable polyamide textiles 100
Cellulose-based synthetic textiles 102
Polylactide textiles 104
Renewable synthetic leather 106
Cotton textiles 108
Wool textiles 110
Leather 112

CHAPTER 3: METALS **114**
Interview: Richard Hutten *116*
Metal recycling **120**
Recycled aluminium for extrusion 122
Recycled aluminium for sheet applications 124
Recycled aluminium for casting 126
Recycled steel 128
Recycled stainless steel 130
Low-carbon metals **132**
Low-carbon aluminium 134
Low-carbon steel 136

CHAPTER 4: CERAMICS AND GLASS **138**
Interview: Ward Massa *140*
Ceramic recycling **144**
Waste-based brick clay 146
Terrazzo 148
Sintered stone 150
Glass recycling **152**
Recycled soda-lime packaging glass 154
Recycled flat and blown glass 156
Recycled fused glass 158
Recycled borosilicate glass 160

CHAPTER 5: WOOD **162**
Interview: Henrik Taudorf Lorensen *164*
Solid wood **168**
Engineered wood **172**
NAF plywood 174
NAF medium-density fibreboard 176
NAF oriented strand board 178
NAF high-pressure laminates 180
Cork composites 182

CHAPTER 6: PAPER **184**
Interview: Riccardo Cavaciocchi *186*
Paper recycling **190**
Recycled packaging paper 192
Recycled paperboard 194
Recycled moulded paper pulp 196
Virgin paper **198**
Packaging paper 200
Barrier paper 202
Translucent paper 204
Alternative-fibre packaging paper 206
Paperboard 208
Barrier paperboard 210
Alternative-fibre paperboard 212
Moulded paper pulp 214

CHAPTER 7: EMERGING SUSTAINABLE MATERIAL TECHNOLOGIES **216**
Interview: Teresa van Dongen *218*
Carbon capture and utilization 222
Natural growth processes 224
Plastic chemical recycling 226
Mixed-plastics mechanical recycling 228
Mono-material plastic composites 230

Notes 232
Glossary 234
Resources 235
Index 236
Acknowledgements and photo credits 240

Foreword

When industrial design was in its infancy, it was centred around drawing a sketch to kick-start the creative process, leaving materials to a later stage. More recently, that starting point has moved towards materials, making much better use of them as a vehicle for storytelling and as a way for products and brands to define user experiences. Increasingly, and against a rapidly accelerating upward trajectory in terms of production, these stories and experiences are becoming focused on sustainability, and materials are one of the main ways that designers have influence when it comes to developing products with reduced environmental impact.

But materials are a challenging starting point because they're in continuous motion and evolution, with a blurred target and endpoint (if such a thing even exists), making them a subject that's incredibly hard to understand fully. I recently heard someone refer to materials and sustainability as the Wild West – a new frontier where we're having to learn and adapt as we find our way.

Add to these complexities and evolving issues the emotive response that we all have towards the environment and you'll see why it can be scary to stick your neck out as a designer – making claims that one material is better for the environment than another while fearing that someone else will unearth a reason as to why your choice doesn't actually yield such a positive contribution. 'Don't kick the kitten', my friend Ed Thomas would tell his design team while he was director of materials design at Nike, meaning that it's okay to put an idea out there that's still forming – to let it see the light of day, even if it may not turn out to be the best solution. Because applying the right material can be such a fraught task, it often makes our choices as designers quite timid.

There are so many opinions in the world as to what's right in terms of material selection, and more often what's wrong. Sustainability will never be as simple as choosing one material over another, and there will never be clear, decisive boundaries between what is good or bad. Sustainability for designers is about taking a position on which approach is best suited to which industry and which product, planning for the optimal lifespan and expected waste stream on disposal to maximize the chances of recovery. Understanding and navigating these twists and turns are the routes that lead to sustainable material selection.

The Wild West of sustainability and materials is a frontier that Daniel has been at the leading edge of in his work as a designer. His deep understanding of, firstly, the role of materials and sustainability in the design and product development process, and secondly, of balancing the functional needs of materials with the intangible emotive values that materials and sustainability have for consumers, has allowed him to gather the most appropriate level of information for this book. Access to this information is essential for designers who are getting to grips with the complexities of materials and sustainability.

There are many books on the environment, but few that look at it specifically through the lens of materials, and in particular, design and materials selection. For this reason, *Better Things*, with its chapters methodically structured around material categories, will guide you through a range of approaches and strategies for effective application. By the end of it, materials and sustainability will have become a less scary and intimidating subject.

Chris Lefteri

Introduction

As far as I'm concerned, product design starts with materials and processes. Without this foundation, there is no product. But despite the central role of materials in product design, it's not easy for designers to find clear and accessible information about their environmental impact. It's not unusual to come across materials and products that are marketed simply as 'plastic free', 'recycled' or 'bio-based' without much actual information about their origins or how to assess their effects on the environment, often leading to confusion and, in the worst cases, greenwashing. Fortunately, a framework for assessing the environmental impact of materials is starting to emerge, backed by consumers' expectations and new legislation in key markets like the European Union.

With this book I have collected all the useful information and data that I've been able to find for a wide range of materials, organized into seven chapters covering plastics, textiles, metals, glass and ceramics, wood, paper, and emerging sustainable material technologies. It aims to give an indication of the strengths and weaknesses of each material from an environmental point of view, as well as an introduction to key sustainability concepts and strategies. Sustainability is a complex topic and it's my hope that this book will clarify current thinking around sustainability, helping you make better decisions about design and material selection.

Measuring environmental impact

Increasingly, material suppliers are expected to measure their environmental impact and make the results publicly available, typically in the form of a cradle-to-gate partial life-cycle analysis (LCA). A cradle-to-gate study covers every step of materials manufacturing, from the extraction and refinement of raw materials, up until the material leaves the factory gates, while a full LCA study covers the entire life cycle of a product, from raw materials and manufacturing to disposal. The diagram shown opposite gives a top-level overview of the scope of the most common LCA variants.

The environmental data that you'll find for the materials in this book is based on cradle-to-gate analyses from suppliers, trade organizations and other sources, referenced

Life-cycle analysis (LCA)

Cradle-to-cradle and cradle-to-grave LCAs cover the entire life of products, while partial LCAs, such as cradle-to-gate and gate-to-gate studies, focus on specific phases, such as materials production and product manufacturing.

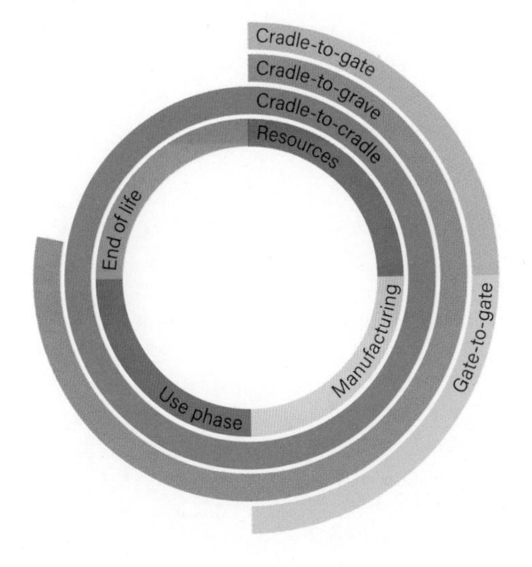

Resources
Extraction and refinement of raw material resources into materials that can be used in products

Manufacturing
Product forming, finishing and assembly, as well as packaging and distribution

Use phase
Energy consumption and other consumables when using the product, as well as maintenance and repairs

End of life
Waste collection and processing, which in a cradle-to-grave life cycle might mean incineration or landfilling, while cradle-to-cradle options would involve product disassembly and recycling, composting or other processes that keep materials in circulation

in the notes at the end of the book (see pages 232–33). It helps to have a basic understanding of the LCA model when looking for environmental information about materials, as suppliers tend to use several different terms – besides cradle-to-gate analysis, these include life-cycle inventory (LCI), environmental product declaration (EPD) and just plain LCA.

I'd like to say at this point that while it's great that a growing number of material suppliers are making this information available, you should think of the numbers that are listed with each material in this book as an indication rather than absolute fact. At the time of writing, LCAs are not always done in the same way, making it difficult to make accurate comparisons of the results between different studies unless you're an expert and know exactly what you're doing. The European Commission has initiated a new methodology and standard for LCAs called the Product Environmental Footprint, or PEF, which is expected to become mandatory in the EU in 2024, and will make it much easier to interpret and compare LCA results. In the meantime, the data in this book should provide guidelines to help you focus your work on problem areas, where you are likely to be able to reduce the environmental impact of the products that you're working on. I would urge you to look both at individual materials as well as the bigger picture, in the form of the statistics and other insights featured at the beginning of each chapter and section of the book. For example, looking at the current situation, you could argue that plastics have more of a waste problem than an emissions problem. By the same token, you could argue that metals have more of an emissions problem than a waste problem. Taking the whole picture into account should make it easier for you to narrow things down and focus on the areas that will make a real difference when making decisions about material selection.

An in-depth introduction to LCA methodology is beyond the scope of this book, but if you're interested in finding out more, the resources on page 235 provide links to LCA software, learning resources and databases with publicly available material environmental data. I'd also urge you to ask suppliers for environmental data about specific materials and grades that you're considering using, as they will often only provide data on request.

Energy, carbon dioxide and materials manufacturing

I think it's reasonable to say that one of the most useful and widely accepted outcomes of LCA studies so far is the ability to calculate greenhouse gas (GHG) emissions and their impact on global warming. Carbon dioxide, or CO_2, is by no means the only GHG, but the volume of CO_2 emissions in materials manufacturing and its impact on global warming makes it very important in the context of this book. To understand the role of carbon in manufacturing and global warming we have to take a closer look at the carbon cycle – a fundamentally important natural process without which life on earth would not be possible.

The carbon cycle consists of two elements – the fast and the slow carbon cycle. The fast carbon cycle connects all life on the planet in real time in a very tangible way. Humans and other animals breathe in oxygen and breathe out CO_2, while plants absorb CO_2 and release oxygen back into the atmosphere. The slow carbon cycle, on the other hand, relies on chemical and tectonic processes that fossilize organic materials and sequester carbon underground and in the ocean floor over the course of hundreds of millions of years.

When we extract and burn large volumes of this sequestered carbon in the form of oil, natural gas or coal to generate energy, it upsets the balance in the fast carbon cycle by releasing more CO_2 into the atmosphere than can be absorbed by the plants on the earth's surface. This is a problem because CO_2, along with other GHGs, traps heat in the atmosphere, slowly warming the planet. The really scary thing about CO_2 is that, if there are not enough plants to absorb it, it remains in the atmosphere forever.

The connection between the materials industry, CO_2 and global warming is at least threefold. Most of the CO_2 emissions from the materials industry come from the energy derived from coal, natural gas and other fossil fuels used in materials production. And materials production often uses a lot of energy. But carbon is also a fundamental building block in many materials – a tree, for example, typically consists of 50 per cent carbon, while the corresponding number for plastics is about 80 per cent. Incinerating these materials releases this embodied carbon into the atmosphere in the form of CO_2. Lastly, overuse of materials that are derived from biomass, such as trees and other plants, can lead to deforestation, reducing the capacity for CO_2 to be absorbed across the globe.

Global warming potential, or GWP, is a measure of how much heat different GHGs trap in the atmosphere, using CO_2 as the benchmark. This way, all GHG emissions can be included in a single number, expressed as CO_2 equivalents, or CO_2e. GHG emissions for the materials featured in this book are all listed as GWP in the form of kilo of CO_2e per kilo of material. However, you should proceed with caution when using these numbers, especially when comparing the impact of different materials. Firstly, a product or part is likely to look very different in terms of design features like wall thickness and reinforcing ribs, as well as overall volume using different materials. Another factor to take into account is that the density of different materials can vary quite a bit. For example, the density of steel is about three times that of aluminium, so comparing the emissions of these materials must be based on the weight of a specific part, rather than kilo for kilo.

Lastly, it's worth pointing out that while GHG emissions are undoubtedly one of the major environmental challenges facing the materials industry, they are by no means the only one. Other aspects of materials manufacturing, such as over-exploitation of renewable and non-renewable resources, toxic ingredients and by-products, and a growing mountain of waste, are also pressing issues in need of sustainable solutions. Let's take a closer look.

The origin of raw materials

At the most basic level, the materials in this book can be categorized as either renewable or non-renewable. Sometimes you'll find that renewable materials are marketed as more sustainable than non-renewable materials by default, although if you dig a little deeper, things may not be so straightforward. Renewable resources like forests can certainly be sustainable, but in reality, irresponsible harvesting of timber is not uncommon, and in extreme cases, illegal logging too, which, if it's allowed to continue, will create enormous environmental issues. As mentioned above, forests are a crucial component in the fast carbon cycle, so felling too many trees will severely impact the global capacity for absorbing CO_2. Over-exploitation of forests will also eventually lead to soil erosion, making it difficult to replant trees (or any other plants, for that matter) and turning the land into desert in the long term. More recently, concerns have been raised about forest monocultures that consist entirely of a single species of tree, which can have a massive impact on local ecosystems and communities. You could also argue that there's an inherent conflict that needs to be carefully handled between commercial forestry, which aims to maximize yield, and natural processes such as allowing dead trees to slowly decay, fulfilling their role as habitats for fungi, insects and other animal life, while also returning nutrients to the soil.

On the other hand, there are non-renewable materials like metals that are among the most widely recycled materials in the world today, to a point where some metal suppliers are preparing for a future where the majority of metal raw materials will come from recycling. Plastics occupy a special place in this context, as it's possible to make them using both non-renewable and renewable raw materials. Plastic production volumes are expected to grow significantly over the coming years, raising major questions about how this increased demand might realistically be met by recycled and renewable plastics.

Other materials are non-renewable, but said to be 'abundant'. Take sand, for example. The kind of sand that's suitable for glass and ceramics production, and used in huge volumes to make cement for the global construction industry, typically comes from riverbeds and beaches, formed as part of a natural cycle in which rocks are gradually eroded by wind and rainwater, with the sediment breaking down into grains of sand as it's washed out into the world's rivers. So while sand is technically a non-renewable resource, it's actually replenished over time. However, the pace of sand mining increasingly exceeds the time it takes for new sediment to replace it, endangering rivers, beaches and sometimes entire coastlines and islands.

Certifications exist for the responsible extraction of many raw materials, which you can read more about in the relevant chapters. Always ask suppliers to confirm the origin of their raw materials and which certifications apply to their materials.

Materials and toxicity

Toxicity is a fairly complex topic, and one that impacts every single stage of material manufacturing. Toxic raw materials and chemical ingredients are used in plastics and synthetic textiles production, for example, although this doesn't automatically mean that the finished material is toxic in itself. Other materials generate toxic by-products during the manufacturing process, such as slag from the mining of metals and ceramic raw materials. Toxic materials are also sometimes used in the refining and processing of raw materials, such as chlorine, which is used for bleaching paper fibre, or carbon disulphide, which is used in the production of viscose textile fibre. Likewise, lead and cadmium were once common in ceramic glazes.

While the above examples are typically subject to tough restrictions and monitoring, or replaced with non-toxic alternatives, there's no shortage of historical examples of industrial disasters involving toxic materials, where the consequences have been devastating for factory workers, local communities and ecosystems. This book lists relevant reduced- or non-toxic options for each material category, as well as individual materials where applicable.

Proper usage of materials is probably of more direct concern to consumers. Many materials are considered safe so long as they're not heated above a certain temperature, or kept out of contact with certain chemicals, water or other substances, for example. There are basic guidelines in place in this area, such as US Food and Drug Administration (FDA) approval and equivalent European legislation, but the bottom line is that we have a fairly primitive understanding of the effects of the complex combinations of chemicals used in materials production. Additives used in plastic materials are often highlighted in this context. While the toxicity of certain additives is well understood – such as the use of bromides for fire resistance, and phthalates for softening PVC materials – there are gaps in our knowledge. A pioneering study published in *Environmental Science & Technology* in 2019 by scientists in Germany and Norway examined a selection of packaging and product samples made with eight common plastic types. While some plastics showed little or no toxicity, others contained several toxic compounds that were not known or documented before the study. This is also a source of concern in terms of recycled plastics, as the plastic recycling processes that are most common at the time of writing are not capable of separating potentially harmful additives from recycled plastic materials. A good starting point is to ensure that any materials, additives and finishes that you are considering are compliant with the EU's **REACH (Registration, Evaluation, Authorisation and Restriction of Chemicals)** and **SVHC (Substances of Very High Concern)** frameworks. I've tried to steer clear of materials with known toxic effects in this book, and my advice would be to keep additives and surface treatments to a bare minimum to avoid complex combinations of materials – toxicity is difficult to predict, given our current limited knowledge.

A return to circularity

Although things may look very different today, what we call circularity has historically been the rule rather than the exception. The economics, infrastructure and technical solutions that make it possible to produce previously unimaginable volumes of products – of which currently only a small proportion are recovered and recycled, composted or put back into the loop in any meaningful way – are thoroughly modern conditions, in stark contrast with a past when resources were more scarce. So in a way, the move towards a more circular economy is going back to the way things were, when reduce, reuse and recycle was the norm rather than jargon. At the same time, it's unreasonable to think that the solution to the environmental crisis is to somehow force the global population back into pre-Industrial Revolution ways of living.

Better Things presents several approaches for dealing with waste recovery and material circularity, and the available options look quite different between different material categories. Recycling is a case in point. Using the processes that we have available today, some materials are easier to recycle than others, which is clearly reflected in the recycling rates of each material category – sometimes down to the level of individual materials and grades. Additionally, recycling is a fairly complex topic in itself. What do we mean by 'recycled material' in the first place? Are we talking about **post-industrial recycled (PIR)** factory waste, potentially collected from the factory floor and put directly back into production, or **post-consumer recycled (PCR)** waste that's been out in the world in a million different places and then recovered, separated from other waste and recycled into new material? The former is usually considerably more straightforward than the latter, which is why PIR and PCR materials should be kept separate when assessing the environmental benefits of recycling. On the other hand, there are many examples of PIR waste that's not easy to recycle, such as mining waste, where recycling would be extremely valuable and desirable.

Another thing to take into account is the quality of the recycled material. Separation of different waste materials is often a prerequisite for any efficient recycling, but several materials also require sorting by specific type and grade of material. Aluminium is a good example,

Global material recycling rates

The ratio of recycled (PIR and PCR) material going into new material production, in million tonnes.[1] A special note about textiles: it's estimated that currently almost all recycled textiles are recycled polyester made with waste PET bottles.

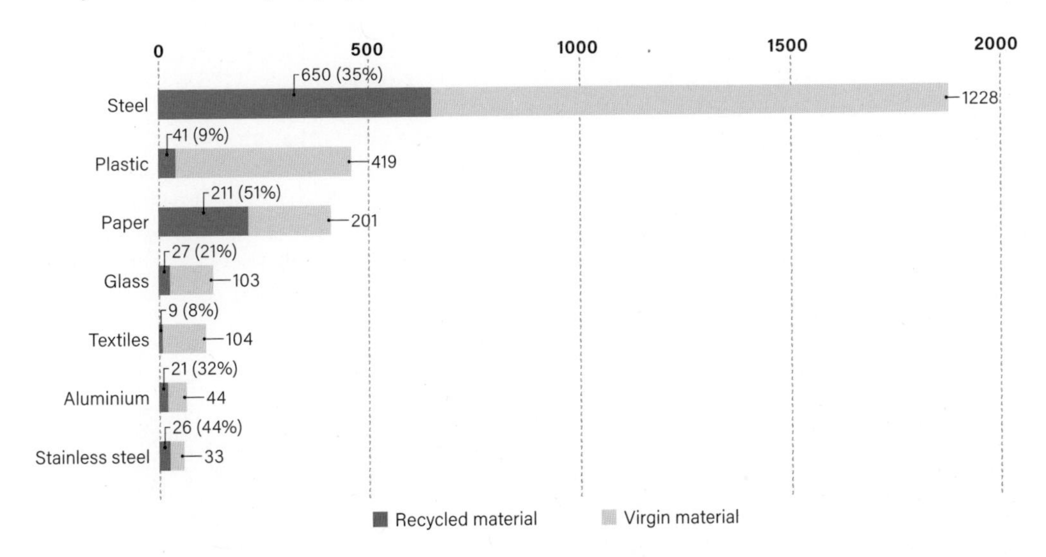

where the mixing of different alloys during recycling will lead to an inferior recycled material. The same is true for plastics, where certain material combinations that are common in packaging applications today make it impossible to recycle the waste into new material with equivalent properties that can go back into packaging applications. While some sorting is done by waste-management companies and recyclers, for the system to work efficiently it's also important that individual households, institutions and companies do their bit. Currently, this is easier said than done, as no universal labelling system is in use across major markets. Many may be familiar with the recycling codes that are sometimes visible on products and packaging, but application is patchy and the system can be seen as too technical and lacking clear direction for how to sort different types of waste for non-experts. Some countries, such as Denmark, have developed their own labelling system, which is used on products, as well as on municipal recycling bins and at recycling centres, to help consumers sort their waste properly. This book provides an introduction to recycling for each featured material category, as well as design approaches for improved circularity.

Beyond recycling, biodegradable materials offer an alternative option at the end of a product's life. Some materials, like paper, are both biodegradable and widely recycled, arguably offering the best of both worlds. There are a few things to take into account with biodegradable materials, however. First of all, there is a difference between biodegradability and compostability, which is often overlooked. 'Biodegradable' simply means that a material will degrade without negative effects on the environment, while 'compostable' describes a material that contains nutrients and other elements that enrich the soil, effectively allowing it to turn into fertilizer during the composting process. There are also significant differences between the conditions that individual materials require to break down. Some need oxygen and elevated temperatures in an industrial composting facility, while others will decompose in a household compost heap or even in landfill. Where applicable, the entries in this book provide an overview of any relevant biodegradability certifications and requirements.

In summary, biodegradable materials should be treated with a degree of caution. Biodegradable waste is capable of causing the same problems as any other type of waste until it's fully broken down, which can take a long time, depending on environmental conditions. For this reason, biodegradable products should not be marketed to consumers in a way that implies that irresponsible littering is somehow acceptable, and well-respected organizations, like TÜV Austria, actively screen applications to avoid awarding certification to products where biodegradability is not helpful or could be seen as advocating guilt-free littering.

Designing for longevity

No material is perfect in every respect; there are always drawbacks that need to be mitigated by making design adjustments or by using combinations of materials or various functional coatings and surface treatments to fulfil specific tasks. Add sustainability into this mix and it quickly becomes very complicated. As mentioned earlier, many common combinations of materials, surface treatments and assembly methods are difficult to separate, which in turn makes recycling difficult, if not impossible. On the other hand, less robust materials can have very negative effects on specific applications through overuse of materials to compensate for lower performance.

One factor to take into account with regards to recycling specifically is that recycled materials often have lower performance than virgin materials, so design adjustments may be required to replace virgin materials with recycled ones. This could mean having to use thicker wall sections and additional features such as reinforcing ribs in plastic mouldings, or using a heavier weight of recycled material in paper applications, to give just two examples.

Each material in this book has key mechanical- and environmental-resistance properties listed, giving an overview of strengths and weaknesses. Solutions are also increasingly being introduced specifically to improve the properties of materials without negatively impacting their circularity, which is highlighted where relevant throughout the book.

The aesthetics of sustainability

One aspect of sustainable materials that's perhaps slightly overlooked is that they often come with significant aesthetic limitations. This can be the result of specific recycling processes, such as plastics, where separation by colour is still rather uncommon; recyclers tend to simply add black or dark grey pigment to give the recycled material colour consistency. But design guidelines for improved circularity can also restrict other aesthetic options – just as some functional finishes and other treatments compromise the circularity of materials, the same applies to decorative finishes. This is of course also a great opportunity to explore new material expressions and user experiences. An overview of relevant aesthetic considerations and compatible finishing processes are provided for all of the materials featured here.

How to use this book

First and foremost, I hope that *Better Things* will be a useful reference, helping you identify ways to reduce the environmental impact of the materials that you specify in your work as a designer. I hope it will also be helpful to you in finding additional materials and suppliers beyond the scope of this book. Knowing roughly where to look will make further research faster and more focused than if simply starting with a generic term like 'sustainable plastics'.

Each entry includes a few suggested suppliers that you can contact as a first step towards sourcing suitable materials. I would like to point out that the selection criteria I've used for suppliers is based on their willingness to make information about the environmental impact of their materials publicly available, as well as my personal view that they are leading in their respective fields. None of the suppliers have been charged to be included in the book, and I have used any information provided by

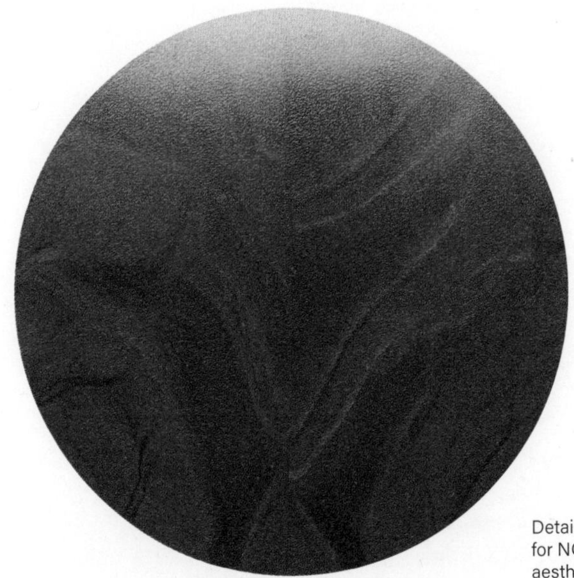

Detail of the S-1500 Chair designed by Snøhetta for NCP (see page 129). On the theme of the aesthetics of sustainability, the colour variations of the recycled plastic waste that's used in the seat are clearly visible in the finished product.

them in good faith, assuming that it's thorough and accurate. Any typos, misinterpretations and other errors are, of course, mine alone. So be sure to follow up and do your own due diligence before entering into any kind of business relationship with suppliers that you find in this book.

On the theme of suppliers, along with statistics and case studies, it's fair to say that this book is rather Eurocentric. This is not a conscious decision on my behalf – my sense is rather that the many European suppliers and case studies in this book are the result of new legislation around sustainability and the European Green Deal initiated by the European Commission, which is forcing material suppliers and brands to reckon with sustainable materials in ways that may not be required yet in other markets. The EU is also a treasure trove for statistics and other data, both through Eurostat, the EU's official statistics agency, but also through the very large number of high-quality reports on everything from waste management to the benefits of renewable raw materials that are the result of projects run by the EU and various business organizations and academic institutions.

Lastly, each chapter in this book features interviews with designers, researchers and entrepreneurs whose work is redefining sustainable design in their respective fields. If the main purpose of *Better Things* is to create an overview of currently available materials, these interviews will open your mind to alternative viewpoints, and give you an idea of how sustainable material technologies may evolve in the future.

1 Plastics

Plastics. Where to start? On the one hand, plastics seem to embody the essence of unsustainable production and use of non-renewable resources; on the other hand, it's almost impossible to imagine the modern world without this family of extremely useful materials. Since the mass production of plastics started in earnest in the early 1900s, the world has seen an explosive growth in plastic products, and as a result, an explosive growth in plastic waste. According to the OECD's 2022 *Global Plastic Outlook* report, some 460 million tonnes of plastic were produced in 2019, and this is expected to reach 1,230 million tonnes by 2060. Given the mountain of plastic waste that we seem unable to deal with today, this paints a rather bleak picture for the future. Another report published in 2016 estimated that the plastics industry represented about 1 per cent of global CO_2 emissions in 2014 – a figure that's expected to reach 15 per cent by 2050. These numbers point to two potential strategies for dealing with the environmental impact of plastics: improved recycling to address plastic waste, and renewable raw materials to address the increasing emissions of the plastics industry.

Breaking down the environmental challenges of these materials, starting with their production, most plastics used today are derived from petrochemicals – in the form of refined crude oil or natural gas. These raw materials are further processed to extract monomers, the molecular building blocks of plastics. Various processes are then applied to these monomers to set off a chain reaction called polymerization, forming the long chains of molecules that make up a polymer (which is just another word for plastic). The complexity of these processes plays a huge role in the energy intensity required in plastics production, and as a result, the emissions generated – plastics that require less processing and fewer manufacturing steps will generate fewer emissions and use fewer resources than those that use more complex processing. Plastics Europe is a trade association for plastics manufacturers, and their set of 'Eco-profiles' (found at **plasticseurope.org**) provides an excellent overview of some of the most common plastic manufacturing processes and their environmental impact.

Plastics – a family tree

An overview of the plastics materials featured in this chapter.

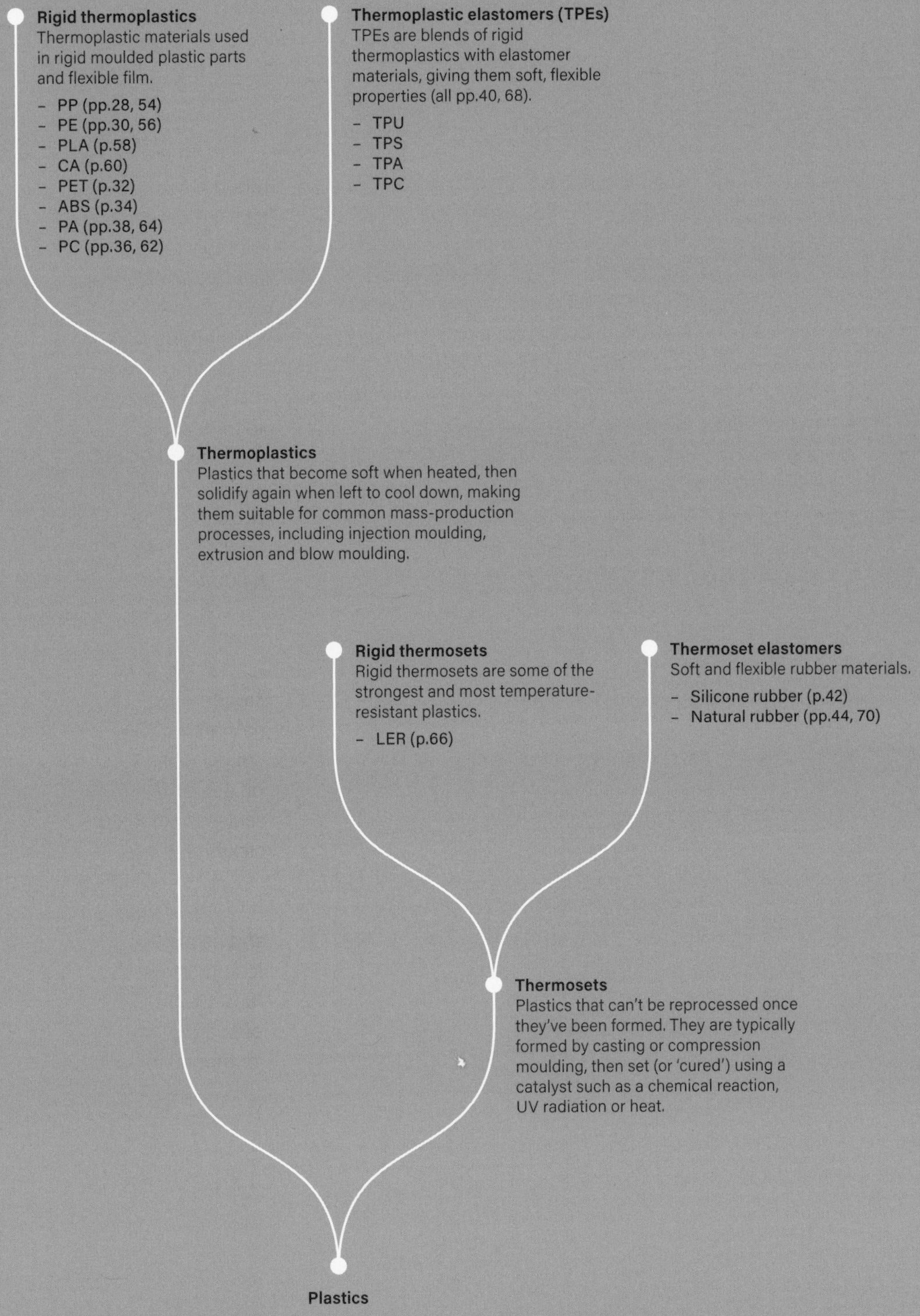

Rigid thermoplastics
Thermoplastic materials used in rigid moulded plastic parts and flexible film.

- PP (pp.28, 54)
- PE (pp.30, 56)
- PLA (p.58)
- CA (p.60)
- PET (p.32)
- ABS (p.34)
- PA (pp.38, 64)
- PC (pp.36, 62)

Thermoplastic elastomers (TPEs)
TPEs are blends of rigid thermoplastics with elastomer materials, giving them soft, flexible properties (all pp.40, 68).

- TPU
- TPS
- TPA
- TPC

Thermoplastics
Plastics that become soft when heated, then solidify again when left to cool down, making them suitable for common mass-production processes, including injection moulding, extrusion and blow moulding.

Rigid thermosets
Rigid thermosets are some of the strongest and most temperature-resistant plastics.

- LER (p.66)

Thermoset elastomers
Soft and flexible rubber materials.

- Silicone rubber (p.42)
- Natural rubber (pp.44, 70)

Thermosets
Plastics that can't be reprocessed once they've been formed. They are typically formed by casting or compression moulding, then set (or 'cured') using a catalyst such as a chemical reaction, UV radiation or heat.

Plastics

Almost all of the plastic materials in common use today fall into one of two categories – **thermoplastics** or **thermosetting plastics**. Thermoplastics are the bigger group. They are typically supplied in pellet form so that they can be easily processed using fast, high-volume processes like **injection moulding**, **blow moulding** and **extrusion**. The forming process is repeatable, making it possible to recycle thermoplastic materials by grinding up thermoplastic waste, reheating and forming it into new parts. By contrast, thermosetting plastic materials (thermosets) are typically provided in a liquid resin form that is set, or cured, with some kind of catalyst, such as a chemical reaction, heat or UV light, using forming processes such as casting and **autoclaving**. Once thermoset resins have been formed, they cannot be heated up and reprocessed, so thermosetting plastics are not easily recycled using the same processes as thermoplastics. (The diagram on page 17 gives an overview of the thermoplastic and thermoset materials featured in this book.)

A great number of additives are commonly used with plastics to provide colour and other aesthetic effects, and to enhance functional performance – improved mechanical properties, weatherability or fire resistance, to name just a few examples. The negative impact of some of these additives is well known, such as brominated flame retardants, perfluorinated chemicals for grease- and waterproofing, and the phthalates used to make plastics like PVC softer. To minimize the use of harmful ingredients, always ask suppliers for compliance with REACH (Registration, Evaluation, Authorisation and Restriction of Chemicals) – a regulation that came into force in the European Union in 2007 – as well as the Restriction of Hazardous Substances Directive (RoHS), which is another set of regulations in the EU specifically for electronic products. But whether these go far enough in identifying potentially harmful additives and ingredients in plastics remains an open question. Independent organizations such as the Stockholm-based non-profit International Pollutants Elimination Network (IPEN) regularly publish reports and findings relating to plastics and toxicity. This is an area that needs much more research, but there are some online resources providing information on chemicals and ingredients, such as the website of the European Chemical Agency (ECHA), and that of its counterpart in the United States, the Chemical Abstracts Service (CAS).

The building blocks of plastics

The ingredients and manufacturing steps of plastics vary greatly, which is directly linked to the environmental impact of the material. The simplified overview below gives two examples: polypropylene (PP), a so-called commodity plastic that is relatively simple to produce, and polyamide (PA), a technical plastic that involves several more steps.

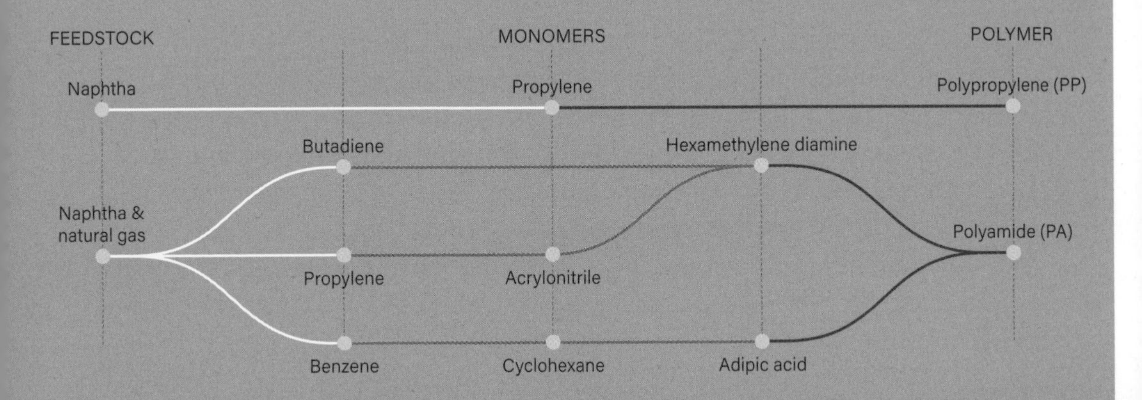

Global warming potential: commodity plastics vs technical plastics

As mentioned above, commodity plastics tend to be relatively simple to produce, while technical plastics are more complex to make, but offer better performance and temperature resistance. As these examples from both categories illustrate, the complexity of the manufacturing process has a direct impact on the global warming potential (GWP) of the resulting material.

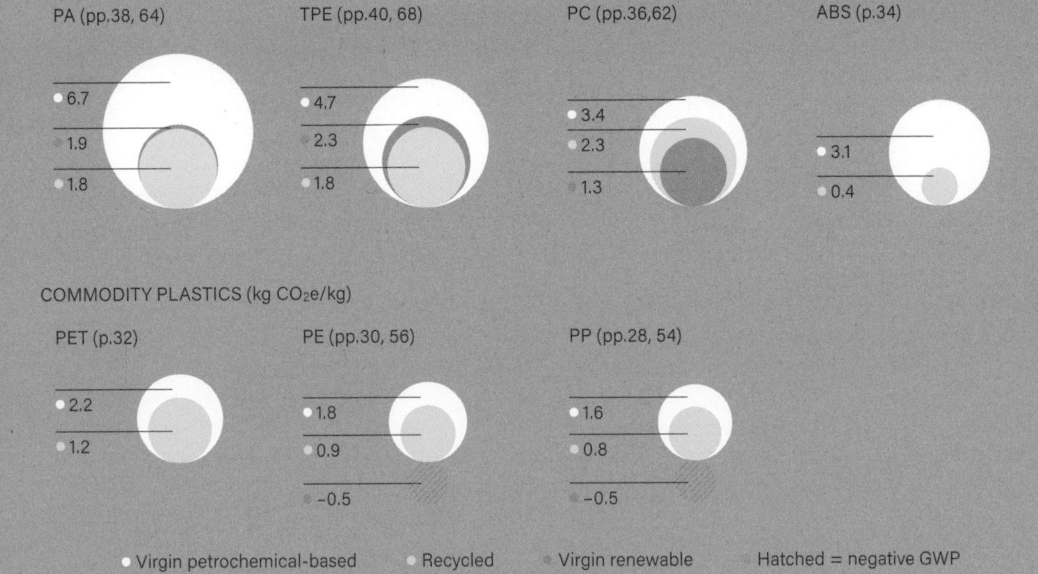

Plastic recycling

The potential environmental benefits of recycled plastics are obvious. Recycling uses less energy than plastics production from virgin raw materials, so results in fewer emissions, and it also keeps plastic waste material out of landfill, incineration plants and the environment. Most plastics in use today are thermoplastics – materials that become soft and malleable when heated, then solidify again when they cool down. The most common plastic recycling process today, so-called mechanical recycling, works in exactly this way.

While this sounds simple enough, in reality mechanical recycling is relatively complex. Somewhat ironically, many of the properties that have made plastics so successful in the first place are what cause problems for recycling. The low cost of commodity plastics, in particular, has a hugely negative impact on plastic recycling, simply because the cost of recycling plastics can be significantly higher than making them from virgin raw materials. Thin and lightweight plastic films may be extremely efficient packaging materials, but they're also flimsy and can be difficult to collect for recycling. Other waste, such as plastic that's been in contact with food or chemicals, may be so contaminated that it's difficult to clean to a sufficient level for recycling into materials that can go back into the loop.

Various incentives are being launched to address these issues, such as proposed new legislation in the European Union (EU) that would require a mandatory recycled material ratio of 30 per cent in plastic packaging by 2030. The hope is that initiatives like this will help drive up plastic recycling rates and turn recycled plastics into a valuable commodity, but for the time being, relatively low volumes of plastic are recycled. According to Eurostat, the EU's statistical office, only 32 per cent of plastic waste was recycled in the EU in 2018. And the United Nations Environmental Programme (UNEP) has estimated that less than 10 per cent of all the plastic that has ever been produced has been recycled. These are depressing numbers, but beyond proposed new rules and legislation there's also lots of room for improvement in terms of our understanding of the strengths and weaknesses of plastics recycling, as well as the impact that design decisions have on the recyclability of plastic products.

Fundamentally, the very large number of different plastics commonly in use today is challenging, as specific plastic materials and grades usually have to be separated to produce high-quality mechanically recycled plastics, avoiding the risk of mixed recycled materials with unpredictable properties and questionable usefulness. A seemingly simple case like the kind of plastic film used in food packaging can consist of several layers of different materials that are difficult, if not impossible, to separate. Many products are complex assemblies that consist of many different parts made with different types of plastic (or other materials altogether), which, if they're joined with adhesives, thermal bonding, **overmoulding** or other processes that make products difficult to take apart, are also very difficult to separate and recycle efficiently.

EU plastic recycling rates

Around 53 million tonnes (Mt) of plastics were used in the EU in 2020, of which around 9 Mt were sent to recycling.[1] After process losses and exports, only about 4.6 Mt of post-consumer recycled plastics found their way into new products, often in completely different industries to those they originally came from.

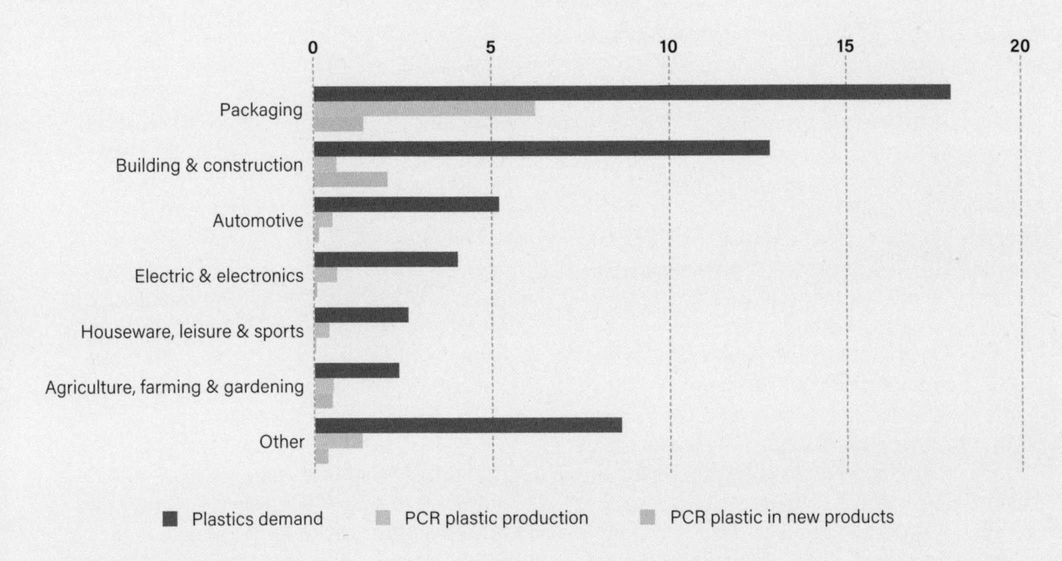

EU plastic recycling capacity

It's estimated that European plastic recyclers have the capacity to recycle 8.5 Mt of plastics on an annual basis. This diagram shows how much of this capacity was allocated to different types of plastics in 2020.[2]

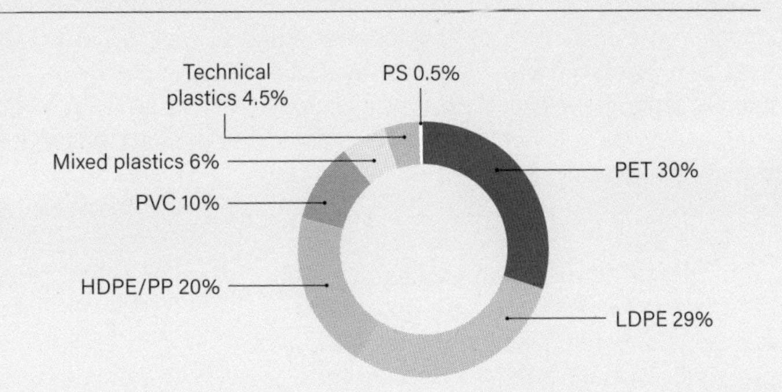

Specific plastics recycling rates

The share of recycled material in relation to the total demand for four specific plastic types where statistics exist, measured in million tonnes.[3]

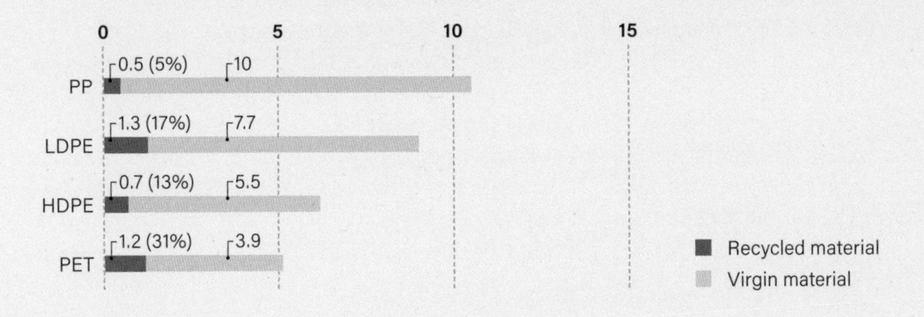

To add to this complexity, the huge variety of additives used in plastic materials can cause other recycling issues. At the most basic level, grinding up and re-melting coloured plastic waste will result in a blend of colours that's unacceptable in most cases where the part is visible in a product. Recyclers typically get around this problem by adding a black or dark grey pigment to create colour consistency, but for light and brightly coloured plastics, much more careful sorting of waste by colour is needed. With transparent plastics, even the slightest contamination will be clearly visible in the final material. Precise sorting and cleaning, of course, require more energy and resources such as water during the recycling process, which is why brightly coloured and transparent recycled plastics have a larger environmental footprint than black or dark grey equivalents.

For other additives, such as the fire retardants that are required in plastics for consumer electronics, for example, closed-loop recycling may be the only workable solution, as additives can contaminate recycled plastic and make it unsuitable for – or outright banned from – sensitive applications such as food packaging or toys.

So to sum up, from a mechanical recycling perspective it's best to try to reduce the number of different plastics in a product, and preferable to make an entire product from a single material where possible. The number of additives should also be reduced to a bare minimum to improve recyclability. And in terms of decoration, branding elements and details that use secondary finishes, such as coatings, films or inserts, should be avoided, unless they can be recycled with the base material without separation.

The good news is that many plastics are available in different grades and qualities, making it possible to create complex products from a single material. The Mono Material collection of jackets and other clothing from the Norwegian brand Helly Hansen, for instance, makes full use of the versatility of polyethylene terephthalate (PET; pages 32–33), a material that's available in textile fibre form, artificial down for insulation, as well as rigid moulded parts like buttons and other fasteners, to give just a few examples. The Cloudneo trainer from the Swiss brand On Running (opposite) is another example – it consists of fewer than ten components, all made with materials from the polyamide family, from the textile in the upper to the soft elastomer in the sole, streamlining the recycling process.

Fortunately, plenty of guidance exists for best practices in this field. The PolyCE project was launched by the European Commission to promote plastics recycling, and its 2021 report *Design for Recycling, Design from Recycling: Practical Guidelines for Designers* is an extremely useful resource for circular design and plastics recycling in consumer electronics applications. Also based in Europe, RecyClass is a plastics industry organization that develops methods for defining the recyclability of plastics and the quality of recycled materials, with a specific focus on packaging.

The global capacity for recycling plastics is expected to grow significantly over the coming years, due to targets and rules relating to mandatory recycled content in plastic products in the EU and other markets. As with any recycled material, always ask suppliers to confirm whether their recycled plastics are made with post-industrial recycled (PIR) or post-consumer recycled (PCR) waste. Most, if not all, standards for assessing the environmental impact of recycled materials make a clear distinction between PIR and PCR materials (see page 12). Beyond mechanical recycling, several alternative recycling technologies for plastics are emerging too, including chemical recycling, which is covered on pages 226–27.

The Cyclon Cloudneo running shoe by On Running. The entire shoe is made with polyamide in different forms, including moulded foam, elastomer and rigid parts, as well as the textile upper, making recycling possible without the need for separation.

Circular plastics

This table gives an overview of plastics that are available in different forms, which can be combined in products and recycled without separation. Adhesives should be avoided unless chemically compatible with the plastic used. Foamed plastics should use gas-assisted foaming rather than chemical foaming agents. Secondary coatings and additives such as restricted pigments, fire retardants, stabilizers and plasticizers should be avoided altogether. For a detailed introduction to circular design guidelines for plastics, see PolyCE's excellent publication *Design for Recycling, Design from Recycling.*

	PA	PP	PET	TPE	PLA	PE	PC	ABS
Rigid opaque	✓	✓	✓	✓	✓	✓	✓	✓
Rigid transparent/translucent	✓	✓	✓	✓	✓	✓	✓	
Film	✓	✓	✓	✓	✓	✓	✓	
Elastomer	✓	✓		✓				
Expanded foam	✓	✓	✓	✓		✓		
Textile fibre	✓	✓	✓	✓	✓			
Self-reinforced composite		✓	✓					
Adhesive	✓							

'There are no sustainable materials – it depends on how you use them.'

An interview with Efrat Friedland, founder of the materials consultancy Materialscout, and co-founder of Positive Plastics, a curated selection of plastic materials based on their reduced environmental footprint compared with conventional plastics. Visit materialscout.com and positiveplastics.eu to find out more.

You've had a long career and extensive experience in materials and design - could you tell me a bit about where you started and how your work has evolved over time?

When I founded my consultancy, Materialscout, the questions and challenges that the team and I usually faced were along the lines of 'Help me find a material that will make my product stand out against the competition', or 'I'm paying this much for materials; help me find less expensive alternatives'. Questions about sustainability were rare at that point, but it's always been a priority for me, and about five years ago we decided to focus exclusively on sustainability and not take on projects that were driven purely by styling or cost from that point onwards.

This started a process where we're constantly educating ourselves and evolving our thinking about sustainability and circular economy. It's easy to think of sustainability as just another materials attribute, but we don't see it that way. As far as I'm concerned, there are no sustainable materials - it depends on where you use them and how you process them. So our work now is very much based around where these questions lead us and our clients, and then applying this thinking in our work.

Materials from Materialscout's sample library.

In addition to Materialscout, you also co-founded and launched Positive Plastics. Given that plastics can be such a fraught topic, I find it very refreshing that you're taking a stand for this complex family of materials.

The materials in the Positive Plastics collection all carry a promise of sustainability. They all have a reduced environmental footprint compared with conventional, petrochemical-based plastics. It's very important for me and the other co-founders of Positive Plastics – Erik Moth-Müller and Markus Paloheimo – that we fully understand where the selected materials come from, how they're produced, what their properties are and what applications they're suitable for.

In terms of recycled plastics, it's mostly about expectations around functionality and aesthetics. Designers should understand that recycled plastics are not identical to virgin materials – there will typically be fewer colours available, and visual imperfections to deal with, such as flow lines. On the functional side of things, tools may need to be adjusted for thicker wall sections to compensate for lower performance, and so on. Some recyclers focus on sorting and cleaning to produce very high-quality recycled materials that can be used as drop-in replacements for virgin plastics. Other recyclers are not as concerned with precise sorting and purity of materials, and as a result, they can offer recycled materials with a smaller environmental footprint.

One of the materials in the Positive Plastics kit is produced by the Israeli supplier UBQ Materials, who created a plastic additive made of unsorted domestic waste – food, paper and cardboard, and other post-consumer plastic waste. This additive is later compounded with various virgin or recycled polymers to reduce the overall material's carbon footprint. The resulting material is unable to compete with engineering plastics like ABS and polyamide, but it can be used in a wide range of applications with lower mechanical requirements. At the other end of the scale, the plastics manufacturer LyondellBasell and the waste-management company SUEZ have launched a joint venture called Quality Circular Polymers (QCP) to produce high-quality materials with a high recycled-content ratio based on careful sorting, enabling a good colour selection and excellent mechanical properties.

It seems to me that plastic recycling is in a very creative phase at the moment. I don't think that the technical potential of different recycling processes is that well understood, and this is resulting in all kinds of experimentation, opening up new possibilities.

Many recyclers offer basic plastic regrind and flakes, but for me the most interesting recyclers are those who are also compounders, meaning they're able to provide materials that are ready to go straight into production. MBA Polymers is one example of this type of company; they have a very interesting portfolio of high-performance plastics, mainly based on post-consumer electronics waste. For me, it boils down to keeping as much plastic waste as possible out of the environment. If this means looking at new hybrid materials or new ways of recycling altogether, such as chemical recycling, then I'm all for it. But we have to know exactly what is in these materials, and their environmental benefits have to be obvious. With mandatory requirements being introduced for recycled plastics in many industries and product categories, we need transparency and traceability. Basically, we need every material to come with a digital passport that carries all this information so that designers, engineers and product managers are able to make informed materials selections.

Materials from the Positive Plastics Kit 2.

DuraPET™ 0624
PCR ST
ReCycle
6050
ALFATER XL® ECO
A75I 2GP0001 BK
GP1015
VDI 30
VDI 36

Recycled polypropylene (PP)

Along with PE and PET (pages 30–31 and 32–33), PP is one of the most widely recycled plastics today, with several suppliers offering it in different grades.

Suppliers & materials	- Fortum Circo® PCR PP - Vogt-Plastic PCR PP - Borealis Borcycle™ PCR PP - LyondellBasell CirculenRecover PCR PP	
Raw material origin	Rigid and flexible packaging waste, automotive exterior and interior parts, furniture and accessories, construction waste. Many recycled PP materials contain a mix of PCR and PIR waste, as well as virgin material. Ask suppliers to provide ISCC PLUS, SCS Recycled Content or other certification that verifies the ratio of recycled content in specific grades.	
	Fortum Circo® PCR PP[4]	**Virgin petrochemical-based PP**[5]
GWP	Ca 0.8 kg CO_2e / kg	1.6 kg CO_2e / kg
Energy use	No data	77.1 MJ / kg
Water use	6.8 l / kg	86.1 l / kg
Toxicity	Confirm the suitability of specific recycled PP grades for food-, water- and skin-contact applications – contamination can occur during the recycling process unless sorting and cleaning is handled to a level that guarantees material quality. Request a material safety data sheet and ask suppliers to confirm that specific materials are REACH compliant.	
Circularity	Recycled PP accounted for an estimated 5% of total demand for PP in Europe in 2018 – about 500,000 tonnes.[6] For an overview of recycling rates and circular design guidelines for plastics, see page 23.	
Mechanical properties	While virgin PP is very versatile, with low weight, good impact resistance and general toughness, recycled PP may have lower performance in some cases. Request a technical data sheet to confirm the performance of specific recycled grades.	
Environmental resistance	Good thermal resistance with an upper **safe service temperature** between 90 and 120°C (195–250°F). Good resistance to moisture and some chemicals, but UV resistance is poor.	
Forming	Compatible with common thermoplastic processes, including injection moulding, blow moulding, extrusion and **thermoforming**. However, recycled PP may need processing and design adjustments to compensate for lower performance compared with virgin material. Request processing guidelines from suppliers.	
Finishing	The majority of recycled PP materials are available in black or shades of dark grey only. Lighter coloured and translucent recycled PP materials are less common, due to the need for careful sorting and cleaning during the recycling process. Most PP grades have poor scratch resistance, but abrasion-resistant grades are available for durable, glossy surfaces.	

Toolbox RE, designed by Arik Levy for Vitra, made with PCR polypropylene supplied by Vogt-Plastic.

vitra.

Recycled polyethylene (PE)

Like PP and PET (pages 28–29, 32–33), PE is one of the most widely recycled plastics today. There are two main varieties: high-density polyethylene (HDPE) and low-density polyethylene (LDPE), both of which are available from several suppliers of recycled PE.

Suppliers & materials	- Fortum Circo® PCR PE - LyondellBasell CirculenRecover PCR PE - SABIC TRUCIRCLE™ PCR PE	
Raw material origin	Flexible and rigid packaging, furniture and industrial applications such as pipes. Many recycled PE materials contain a mix of PCR and PIR waste, as well as virgin material. Ask suppliers to provide ISCC PLUS, SCS Recycled Content or other certification that verifies the ratio of recycled content in specific grades.	
	Fortum Circo® PCR HDPE[7]	**Virgin petrochemical-based HDPE**[8]
GWP	0.9 kg CO_2e / kg	1.8 kg CO_2e / kg
Energy use	No data	79.3 MJ / kg
Water use	7.3 l / kg	105.5 l / kg
Toxicity	Confirm the suitability of specific recycled PE grades for food-, water- and skin-contact applications – contamination can occur during the recycling process unless sorting and cleaning is handled to a level that guarantees material quality. Request a material safety data sheet and ask suppliers to confirm that specific materials are REACH compliant.	
Circularity	Recycled HDPE accounted for an estimated 14% of total demand for HDPE in Europe in 2018 – about 700,000 tonnes.[9] For an overview of recycling rates and circular design guidelines for plastics, see page 23.	
Mechanical properties	Virgin HDPE has good rigidity and tensile strength, while LDPE is more flexible and suitable for films, but recycled PE grades may have lower performance in some cases. Request a technical data sheet to confirm the performance of specific recycled grades.	
Environmental resistance	Good thermal and moisture resistance, as well as to some chemicals, but UV resistance is poor.	
Forming	Compatible with common thermoplastic processes, including injection moulding, blow moulding, extrusion and thermoforming. However, recycled PE may need processing and design adjustments to compensate for lower performance compared with virgin material. Request processing guidelines from suppliers.	
Finishing	The majority of recycled PE materials are available in black or shades of dark grey only. Lighter coloured recycled PE materials are available, but to a lesser extent, due to the need for careful sorting and cleaning during the recycling process. PE has rather poor scratch resistance, making it unsuitable for durable, glossy surfaces.	

Butter Chair, designed by Sarah Gibson and Nicholas Karlovasitis (DesignByThem). Made from 100% recycled PCR HDPE, the Butter collection is UV stable, waterproof and available in a range of colours. To reduce its freight footprint, the design starts as flat sheets of material that are hand-folded to become rigid.

Recycled polyethylene terephthalate (PET)

The large quantities of transparent PET waste available from bottles make this an attractive material for recycling into new transparent products, or for colour flexibility, as pigments can easily be added to clear material.

Suppliers & materials	- Indorama Deja™ PCR PET - Veolia CleanPET® PCR PET	
Raw material origin	Drink bottles, packaging trays and blister packs, as well as automotive and industrial parts. Many recycled PET materials contain a mix of PCR and PIR waste, as well as virgin material. Ask suppliers to provide ISCC PLUS, SCS Recycled Content or other certification that verifies the ratio of recycled content in specific grades.	
	Indorama Deja™ PCR PET (pellets)[10]	**Virgin petrochemical-based PET**[11]
GWP	1.2 kg CO_2e / kg	2.2 kg CO_2e / kg
Energy use	No data	69.6 MJ / kg
Water use	No data	95.8 l / kg
Toxicity	Confirm the suitability of specific recycled PET grades for food-, water- and skin-contact applications –contamination can occur during the recycling process unless sorting and cleaning is handled to a level that guarantees material quality. Request a material safety data sheet and ask suppliers to confirm that specific materials are REACH compliant.	
Circularity	Recycled PET accounted for an estimated 26% of the total amount of PET produced in Europe in 2021.[12] Beyond plastics, recycled PET can also go into polyester textile production, as detailed on pages 80–81. For an overview of recycling rates and circular design guidelines for plastics, see page 23.	
Mechanical properties	While virgin PET has good all-round mechanical properties with a good strength-to-weight ratio and impact resistance, recycled PET may have lower performance in some cases. Request a technical data sheet to confirm the performance of specific recycled grades.	
Environmental resistance	Good resistance to chemicals, moisture and UV radiation, but doesn't perform as well under high temperatures, especially when in contact with boiling water.	
Forming	Beyond blow-moulded bottles and thermoformed trays, PET can be challenging to form using injection moulding. PET-forming specialists should be able to help with specific requirements.	
Finishing	Much PET waste comes from clear bottles, making it easier to produce light coloured and clear recycled PET compared with many other plastic waste streams, which are likely to be a jumble of opaque colours. However, transparent recycled PET still requires rather intensive cleaning, which adds to its cost and environmental footprint. PET has good scratch resistance, making it suitable for polished, glossy surfaces.	

Kuggis storage box by IKEA.
The box is made with recycled PET waste.

Recycled acrylonitrile butadiene styrene (ABS)

All-round high performance at a relatively low cost has made ABS one of the most common engineering polymers. It's also one of the most widely recycled materials in this family, with several suppliers offering high-quality recycled grades.

Suppliers & materials	- MBA Polymers PCR ABS - INEOS Styrolution Terluran® ECO PCR ABS - INEOS Styrolution Novodur® PCR ABS	
Raw material origin	Consumer electronics, appliances, automotive parts and toys, to name just a few. Many recycled ABS materials contain a mix of PCR and PIR waste, as well as virgin material. Ask suppliers to provide ISCC PLUS, SCS Recycled Content or other certification that verifies the ratio of recycled content in specific grades.	
	MBA Polymers PCR ABS[13]	**Virgin petrochemical-based ABS**[14]
GWP	0.4 kg CO_2e / kg	3.1 kg CO_2e / kg
Energy use	No data	90.6 MJ / kg
Water use	No data	22 l / kg
Toxicity	Confirm the suitability of specific recycled ABS grades for food-, water- and skin-contact applications – contamination can occur during the recycling process unless sorting and cleaning is handled to a level that guarantees material quality. Recycled ABS may also contain flame-retardant additives and will give off toxic fumes when heated above 200°C (390°F). Request a material safety data sheet and ask suppliers to confirm that specific materials are REACH compliant.	
Circularity	Although statistics were unavailable at the time of writing, it's reasonable to assume that recycled ABS is one of the more widely recycled technical plastic materials, based on its relatively high availability in the market. For an overview of recycling rates and circular design guidelines for plastics, see page 23.	
Mechanical properties	While virgin ABS has good all-round mechanical performance, with excellent impact resistance, rigidity and scratch resistance, recycled ABS may have lower performance in some cases. Request a technical data sheet to confirm the performance of specific grades.	
Environmental resistance	Good resistance to some chemicals, coupled with good UV, moisture and thermal resistance.	
Forming	Compatible with common thermoplastic processes, including injection moulding, extrusion and thermoforming. However, recycled ABS may need processing and design adjustments to compensate for lower performance compared with virgin material. Request processing guidelines from suppliers.	
Finishing	Most recycled ABS is available in black or shades of dark grey only. Lighter coloured recycled ABS materials are less common, due to the need for careful sorting and cleaning during the recycling process. ABS has good scratch resistance, making it suitable for polished, glossy surfaces.	

Pure D9 Green vacuum cleaner by Electrolux. The exterior is made with 70% recycled ABS.

S-BAG
Electrolux
PURED9
S-BAG
MIN
MAX
GREEN

Recycled polycarbonate (PC)

Like most technical plastics, PC is currently not widely recycled, but it's attractive in that much PC waste is transparent, making it easier to produce light-coloured and transparent recycled materials.

Suppliers & materials	- Covestro Makrolon® PCR PC - TRINSEO™ EMERGE™ ECO PCR PC	
Raw material origin	Furniture and accessories, consumer electronics, automotive parts and architectural glazing, to name just a few. Many recycled PC materials contain a mix of PCR and PIR waste, as well as virgin material. Ask suppliers to provide ISCC PLUS, SCS Recycled Content or other certification that verifies the ratio of recycled content in specific grades.	
	Covestro Makrolon® QC 50% PCR PC[15]	**Virgin petrochemical-based PC**[16]
GWP	2.1 kg CO_2e / kg	3.4 kg CO_2e / kg
Energy use	No data	99 MJ / kg
Water use	No data	1,535 l / kg
Toxicity	Bisphenol A (BPA), a key chemical ingredient in PC, has possible links to hormone disruption in humans and animals, so PC should be avoided for food- and water-contact applications. Additionally, because PC is often used in applications that require fire resistance, it may contain flame-retardant additives. Request a material safety data sheet and ask suppliers to confirm that specific materials are REACH compliant.	
Circularity	Statistics were unavailable at the time of writing, but it's probably safe to assume that PC is not currently widely recycled, based on the relatively low availability of recycled PC in the market at the time of writing. For an overview of recycling rates and circular design guidelines for plastics, see page 23.	
Mechanical properties	While virgin PC is a lightweight, tough material with good impact and abrasion resistance, recycled PC may have lower performance in some cases. Request a technical data sheet to confirm the performance of specific grades.	
Environmental resistance	Good UV and thermal resistance, as well as good resistance to certain chemicals.	
Forming	Compatible with common thermoplastic processes, including injection moulding, extrusion and thermoforming. However, recycled PC may need processing and design adjustments to compensate for lower performance compared with virgin material. Request processing guidelines from suppliers.	
Finishing	Most recycled PC is available in black or shades of dark grey only, but given the relatively large quantities of clear PC waste from transparent applications such as lighting and glazing, several suppliers are able to offer light coloured and transparent or translucent recycled PC. PC has good scratch resistance, making it suitable for polished, glossy surfaces.	

The rear cover of the Fairphone 4 uses Makrolon® PCR polycarbonate from Covestro, made with waste from automotive headlights, CDs, architectural sheet materials and other sources.

FAIRPHONE

Recycled polyamide (PA)

Tough and durable PA (also called nylon) is among the more widely recycled engineering polymers. The relatively high environmental footprint of virgin PA makes recycled PA an attractive option, available from several suppliers.

Suppliers & materials	- Aquafil ECONYL® PCR PA6 - DSM Akulon® RePurposed PCR PA6 - Arkema Virtucycle® PCR PA11	
Raw material origin	Automotive and industrial parts, furniture, appliances and sporting equipment. However, many recycled PA materials contain a mix of PCR and PIR waste, as well as virgin material. Ask suppliers to provide ISCC PLUS, SCS Recycled Content or other certification that verifies the ratio of recycled content in specific grades.	
	Aquafil ECONYL® PCR PA6[17]	**Virgin petrochemical-based PA6**[18]
GWP	1.8 kg CO_2e / kg	6.7 kg CO_2e / kg
Energy use	55.4 MJ / kg	128.8 MJ / kg
Water use	2,550 l / kg	1,647 l / kg
Toxicity	Confirm the suitability of specific recycled PA grades for food-, water- and skin-contact applications – contamination can occur during recycling unless sorting and cleaning is handled to a level that guarantees material quality. PA is often used in applications requiring fire resistance, such as consumer electronics and automotive parts, so recycled PA may contain flame-retardant additives. It will also give off toxic fumes above 300°C (570°F). Request a material safety data sheet and confirm that materials are REACH compliant.	
Circularity	Statistics were unavailable at the time of writing, but recycled PA is likely one of the more widely recycled engineering polymers, based on relatively high availability. Beyond plastics, recycled PA can also go into PA textile production (see pages 82–83).For an overview of recycling rates and circular design guidelines for plastics, see page 23.	
Mechanical properties	While virgin PA has excellent impact strength and stiffness, as well as good abrasion and wear resistance, recycled PA may have lower performance in some cases. Request technical data sheets to confirm mechanical properties.	
Environmental resistance	Very good chemical, UV and thermal resistance. Some types of PA have poor moisture resistance, however, notably PA6 and 6.6.	
Forming	Generally straightforward to form using common thermoplastic processes, including injection moulding, extrusion and thermoforming, but moulded PA6 and 6.6 parts can be susceptible to warping and other distortion due to poor moisture resistance. Recycled PA may need processing and design adjustments to compensate for lower performance compared with virgin material. Request processing guidelines from suppliers.	
Finishing	Most recycled PA is limited to black or shades of dark grey. Lighter coloured and transparent recycled materials are less common due to the need for careful sorting and cleaning during the recycling process, which also adds to their cost and environmental footprint. PA has good scratch resistance, making it suitable for polished, glossy surfaces.	

Zettle Ocean Reader credit card reader. The front cover uses recycled polyamide made with marine waste such as fishing nets and ropes.

€25,00
Tap/Insert Card
Zettle

Recycled thermoplastic elastomers (TPEs)

Soft and flexible TPEs are not currently widely recycled, possibly because they are often used in overmoulded parts in combination with other materials, making them difficult to separate for recycling.

Suppliers & materials	- TRINSEO™ APILON™ ECO PIR TPU (thermoplastic polyurethane) - Covestro Desmopan® PCR TPU (thermoplastic polyurethane) - HEXPOL Dryflex® Circular PCR TPS (thermoplastic styrenic block copolymer) - Arkema Virtucycle® Pebax® PCR TPA (thermoplastic polyether block amide)	
Raw material origin	Footwear, tool handles, automotive parts and household products. However, the recycled TPEs featured here contain a mix of PCR and PIR waste, as well as virgin material. Ask suppliers to provide ISCC PLUS, SCS Recycled Content or other certification that verifies the ratio of recycled content in specific grades.	
	TRINSEO™ APILON™ ECO PIR TPU[19]	**Virgin petrochemical-based TPU**[20]
GWP	1.8 kg CO_2e / kg	4.7 kg CO_2e / kg
Energy use	55.4 MJ / kg	50 MJ / kg
Water use	2,550 l / kg	59 l / kg
Toxicity	Confirm the suitability of specific recycled TPE grades for food-, water- and skin-contact applications – contamination can occur during the recycling process unless sorting and cleaning is handled to a level that guarantees material quality. Request a material safety data sheet and confirm that materials are REACH compliant.	
Circularity	TPEs are not widely recycled, possibly because of low volumes, but also because they're often used in combination with other materials, so, to maximize the likelihood of recycling, try to make TPE parts easily separable, or make the entire product from TPE if possible. Sometimes specific TPEs can be recycled with compatible materials without separation. For an overview, see page 23.	
Mechanical properties	All TPEs can be stretched without permanent deformation. Several grades are typically available with different flexibility and softness properties. Tough, with good tear and abrasion resistance, but recycled TPEs may have lower performance than virgin material. Request technical data sheets for specific recycled grades from suppliers.	
Environmental resistance	Most TPEs have good temperature resistance, but UV, moisture and chemical resistance varies. Confirm the suitability of specific TPE materials for your application with suppliers.	
Forming	Compatible with common thermoplastic processes such as injection moulding and extrusion. However, recycled TPEs may need processing and design adjustments due to lower performance compared with virgin material. Request processing guidelines from suppliers.	
Finishing	Most recycled TPEs are limited to black or dark grey. Lighter coloured and translucent recycled TPE is less common, due to the need for careful sorting and cleaning during the recycling process. TPEs are warm and soft to the touch, with good potential for in-mould textures for tactile effects.	

Fairphone 4 protective cover made with Desmopan® CQ PIR TPU from Covestro. Desmopan® CQ is made with translucent TPU waste, making it easier to add colour to the recycled material. (See also pages 36–37.)

FAIRPHONE

Recycled silicone rubber

Recycled silicone rubber is available from a few specialist suppliers in the form of ground-up silicone rubber waste mixed with virgin silicone, giving the material a distinctive speckled look.

Suppliers & materials	- Adpol Silicrumb PCR silicone - ECO USA PCR silicone oil for silicone production - On-site waste recycling at manufacturing lines	
Raw material origin	PIR silicone waste includes excess material from manufacturing and construction waste from building sites, while PCR waste may include electronics accessories, kitchenware, appliances and automotive parts.	
	Recycled silicone[21]	**Virgin silicone**[22]
GWP	1 kg CO_2e / kg	7.1 kg CO_2e / kg
Energy use	No data	86 MJ / kg
Water use	No data	210 l / kg
Toxicity	While virgin silicone rubber is considered non-toxic and suitable for food-, water- and skin-contact applications, as well as demanding medical applications, the suitability of specific recycled silicone rubber grades for these applications should be confirmed. Recycled silicone can be contaminated unless sorting and cleaning is handled to a level that guarantees material quality. Request a material safety data sheet and confirm that materials are REACH compliant.	
Circularity	Silicone rubber is not widely recycled, possibly because of low volumes, but also because it's often used in overmoulded parts in combination with other materials or as a coating that's difficult to separate for recycling. Try to make silicone parts easily separable, or make the entire product from silicone if possible. For an overview of recycling rates and circular design guidelines for plastics, see page 23.	
Mechanical properties	Very tough and elastic, with excellent tear strength and elongation properties. However, recycled silicone rubber may have lower performance in certain areas compared with virgin material. Request technical data sheets for specific recycled grades from suppliers.	
Environmental resistance	Exceptional thermal resistance, capable of withstanding temperatures of between -50°C and 350°C (-60° to 660°F). Silicone rubber is virtually unaffected by UV radiation and is resistant to most chemicals.	
Forming	Casting and **compression moulding** are common forming processes, but injection moulding is also possible on special machines that allow for curing of the material during the process. However, recycled silicone rubber may need processing and design adjustments to compensate for lower performance compared with virgin material. Request processing guidelines from suppliers.	
Finishing	The colour of recycled silicone rubber will depend on the colour of the recycled waste. Recycled silicone rubber particles will be clearly visible in the material surface. The colour and transparency of the virgin silicone rubber that's mixed with recycled particles can be specified independently. The size of the particles will have an impact on tactility - coarser particles will give a grippy, rough surface, while finer particles will give a smoother surface.	

Aquapulse Pro swimming goggles by Speedo. The visible particles in the band are ground up PIR silicone rubber waste from Speedo's own manufacturing lines, mixed with virgin material and put into new products.

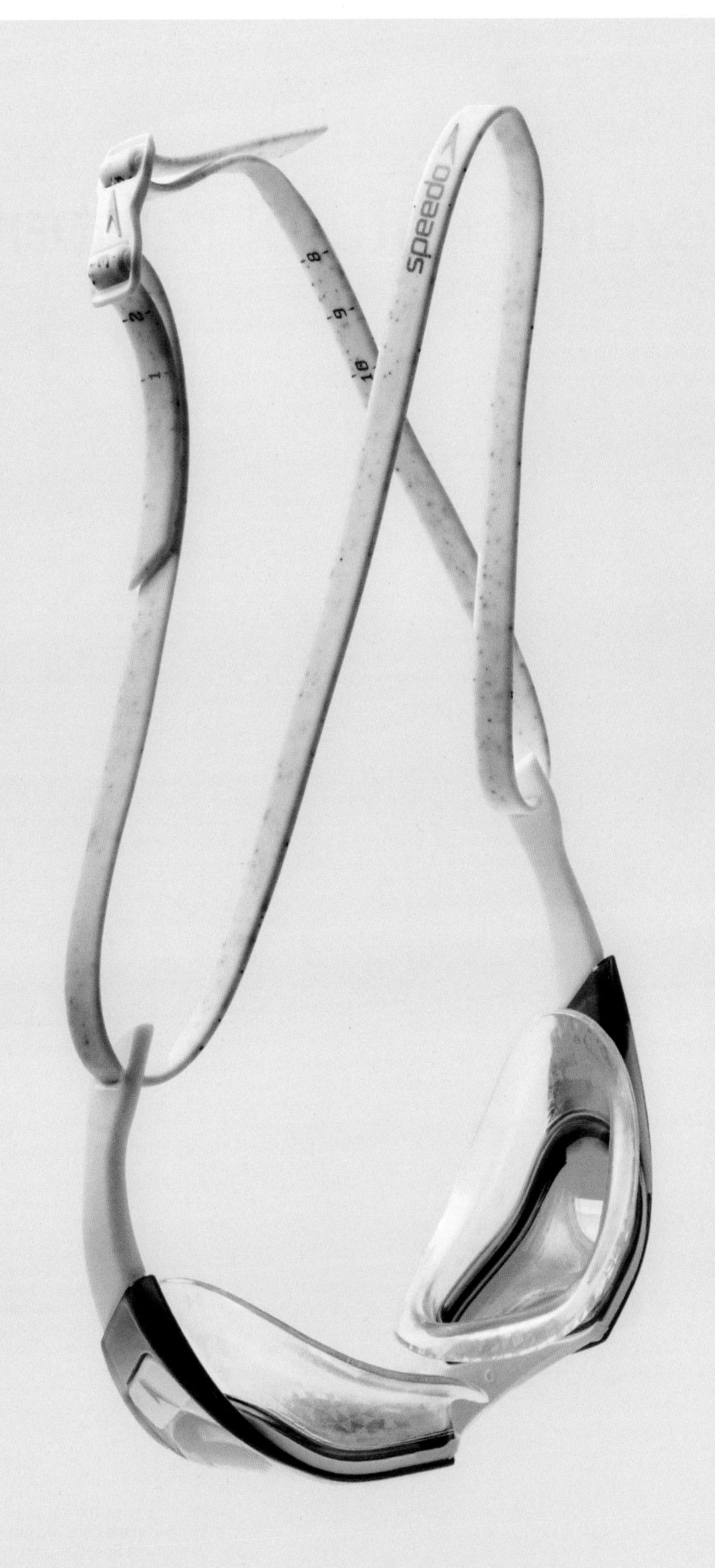
speedo

Recycled natural rubber

Natural rubber is fairly widely recycled, but its usefulness depends on the recycling process: either the material is ground into powder and used as a filler in composite materials, or it's broken down chemically (so-called 'devulcanization'), resulting in a material with like-for-like properties with virgin natural rubber.

Suppliers & materials	- Rubberlink devulcanized PCR rubber - Maris Evorec devulcanized PCR rubber - GEZOLAN PIR rubber granulate	
Raw material origin	Used tyres, conveyor belts, protective bumpers, footwear and protective gloves. Recycled natural rubber is often made with a mix of recycled and virgin material, so ask suppliers to provide ISCC PLUS, SCS Recycled Content or other certification that verifies the ratio of recycled content in specific grades.	
	PCR natural rubber granulate[23]	**Virgin natural rubber (vulcanized)**[24]
GWP	-0.7 kg CO_2e / kg	2.5 kg CO_2e / kg
Energy use	No data	42 MJ / kg
Water use	No data	10 l / kg
Toxicity	Confirm the suitability of specific grades for food-, water- and skin-contact applications - contamination can occur during recycling unless sorting and cleaning is handled to a level that guarantees material quality. Natural rubber and some of the additives used during **vulcanization** can cause serious allergic reactions. Request a material safety data sheet and confirm that materials are REACH compliant.	
Circularity	Around half of all used rubber tyres in Europe are recycled currently;[25] most are ground into granulate, with a smaller amount going to devulcanization. Natural rubber is often used in combination with other materials, so, to maximize the likelihood of recycling, try to make natural rubber parts easily separable, or make the entire product from natural rubber if possible. For an overview of recycling rates and circular design guidelines for plastics, see page 23.	
Mechanical properties	Natural rubber is tough and flexible, with excellent abrasion and tear resistance. However, recycled natural rubber may have lower performance in certain areas compared with virgin material. Request technical data sheets for specific recycled grades.	
Environmental resistance	Good resistance to water and certain chemicals, but not as effective at resisting UV radiation or temperatures above 80°C (180°F).	
Forming	Common forming processes include casting, **dip moulding** and compression moulding. However, recycled natural rubber may need processing and design adjustments to compensate for lower performance compared with virgin material. Request processing guidelines from suppliers.	
Finishing	Since tyres are a major waste stream, much recycled rubber is black. Other natural rubber waste offers wider colour and translucency options in recycling, but these materials are less widely available due to lower volumes and the need for careful sorting and cleaning during recycling. Recycled natural rubber particles may be visible in the material surface, and the size of particles will have an impact on tactility - coarser particles will give a grippy, rough surface, while finer particles will give a smoother surface.	

1L11-01 trainer by Norm. The outsole is made with a mix of 70% PCR natural rubber and 30% virgin fair-trade rubber.

Renewable plastics

For all the hype surrounding renewable plastics, this is a very complex family of materials, and tough questions need to be asked about the origins of plant-based raw materials, their environmental footprint and impact on circularity.

Many different types of plant-based raw material – or biomass, as it's often called in the industry – can be used in renewable plastics, including food crops and other agricultural products, as well as plants that are grown on non-arable land (such as trees) or even at sea (such as algae). All plants absorb CO_2 while they grow and this embodied CO_2 counts towards the carbon footprint of renewable materials, sometimes giving a negative carbon footprint, if the amount of CO_2 sequestered inside the material is greater than the emissions generated during production.

Replacing a conventional petrochemical-based plastic with a renewable alternative should always be considered in context. Renewable plastics typically have a lower carbon footprint than petrochemical-based materials, but other parameters – such as water use, and whether production relies on crops that are in direct competition with the food industry – should also be taken into account. Additionally, many raw materials used in renewable plastics production have their own environmental issues, such as plants that are grown using pesticides and fertilizers to grow. Numerous certifications and initiatives exist to ensure sustainable agricultural practices for material production, such as Bonsucro for sustainable sugar, a common raw material in renewable plastics, as well as certification for non-agricultural raw materials, such as sustainable forest products from the Forest Stewardship Council (FSC) and the Programme for the Endorsement of Forest Certification (PEFC).

The diagram shown opposite gives an overview of the plant-based raw materials used in the production of the renewable plastic materials featured in this book.

As mentioned earlier, the basic ingredients of petrochemical-based plastics are hydrocarbon **feedstocks** such as naphtha, derived from crude oil or natural gas. A growing number of suppliers are developing processes for producing renewable feedstocks, including the Finnish forest-industry company UPM. Since 2015, UPM has produced renewable fuels and feedstocks derived from so-called crude tall oil, a by-product of the paper-pulping process that takes place at their paper mills, using sustainable timber. Their renewable BioVerno naphtha can be used as drop-in replacement for petrochemical-based naphtha.

Other suppliers have developed processes that differ from those used in the petrochemical world. The Brazilian supplier Braskem, for example, uses fermentation to produce renewable polyethylene with sugar derived from sugarcane. Unlike these materials from UPM and Braskem, which can be used as drop-in replacements for petrochemical-based naphtha or polymers, other processes for making renewable plastics have led to the development of new plastic materials altogether. Luminy® PLA from the Netherlands-based supplier Total Corbion is one such example. PLA, which stands for polylactide, is a plastic material in the polyester family that can be derived from a wide range of

Renewable plastic materials

The renewable plastics featured in this book, grouped according to the plant-based raw materials used in their production. As shown, some materials can be derived from several different sources.

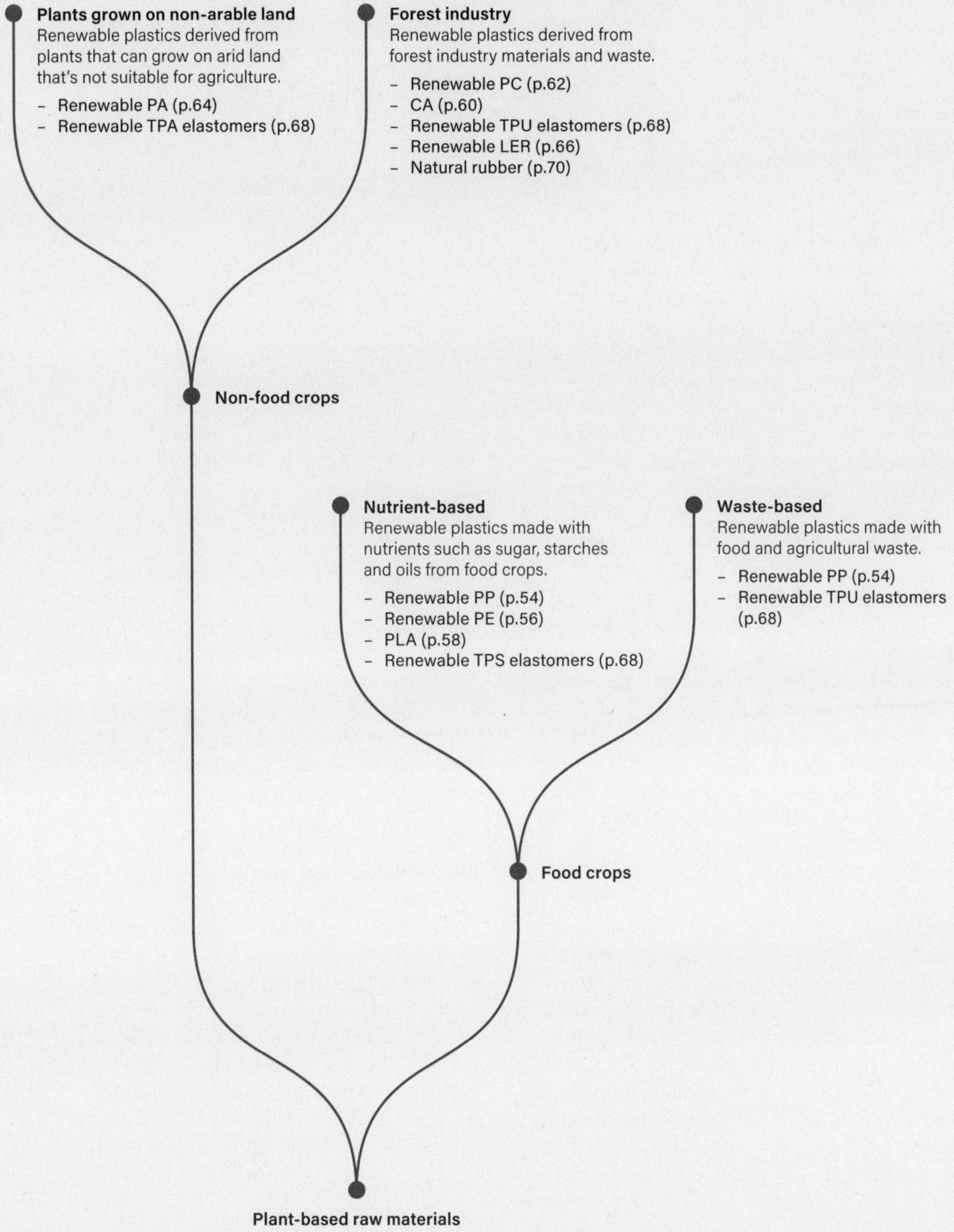

Plants grown on non-arable land
Renewable plastics derived from plants that can grow on arid land that's not suitable for agriculture.

- Renewable PA (p.64)
- Renewable TPA elastomers (p.68)

Forest industry
Renewable plastics derived from forest industry materials and waste.

- Renewable PC (p.62)
- CA (p.60)
- Renewable TPU elastomers (p.68)
- Renewable LER (p.66)
- Natural rubber (p.70)

Non-food crops

Nutrient-based
Renewable plastics made with nutrients such as sugar, starches and oils from food crops.

- Renewable PP (p.54)
- Renewable PE (p.56)
- PLA (p.58)
- Renewable TPS elastomers (p.68)

Waste-based
Renewable plastics made with food and agricultural waste.

- Renewable PP (p.54)
- Renewable TPU elastomers (p.68)

Food crops

Plant-based raw materials

sugars, including European sugar beets and Thai sugarcane. The sugar is fermented and further processed into PLA at Total Corbion's production site in Thailand. Other renewable plastics require hardly any processing at all; natural rubber, for instance, can be tapped from rubber trees and used straight away as a material for simple casting, or for waterproofing textiles, for example.

Many plastics casually described as 'renewable' are in fact only partially made with renewable raw materials. A few approaches and certifications are used to measure and confirm the renewable content of plastic materials – such as OK Biobased from TÜV Austria, which clearly states a fixed percentage of renewable content in materials and products – but this type of certification is not suitable for all suppliers, especially those that make both renewable and petrochemical-based plastics at the same production sites. In these cases, renewable and petrochemical-based feedstocks are typically mixed during production, so the ratio of renewable raw materials varies over time. At the time of writing, the most common way to deal with this scenario is probably the 'mass balance' approach, for which ISCC PLUS is a well-established standard. Essentially a bookkeeping system, ISCC PLUS is a method for tracking and certifying the average ratio of renewable content during a specified period.

Further adding to their complexity, renewable plastics are often assumed to be biodegradable. This is far from the case, as can be seen in the table opposite, which gives an overview of biodegradability in relation to the renewable plastics featured in this book. There are great differences in the conditions that biodegradable plastics require to degrade within a reasonable time frame. Some will degrade in a garden compost heap, while others require industrial composting facilities. Several certifications exist in this area, with TÜV Austria's OK Compost and OK Biodegradable being two common examples. These distinguish between compostable materials, which effectively yield organic fertilizer, and biodegradable materials, which simply degrade without adverse effects on the environment.

Although these certifications and others like them are widely used, the use of the term 'biodegradability' in product marketing is controversial – it can be seen as sending a message to consumers that it's 'safe' to dispose of biodegradable materials in the environment. Until biodegradable plastics have fully degraded, they're indistinguishable from other plastic litter and they're capable of causing the same environmental problems. As a result, most organizations that certify biodegradability will assess applications individually to discourage frivolous marketing and questionable product claims. It's also worth noting that all biodegradable and compostable materials emit greenhouse gases during the degradation process.

To sum up, the same design guidelines that apply to plastic materials in general also apply to renewable plastics. Like their petrochemical-based counterparts, renewable plastics – including biodegradable varieties – generate waste. It's also important to consider the impact of new types of plastic, such as PLA, which are not widely recycled today, while others, like renewable polypropylene, polyethylene and polyamide, can be recycled together with their petrochemical-based counterparts in established waste streams. Lastly, always ask suppliers of renewable plastics for documentation that confirms raw materials have been grown in accordance with sustainable agricultural and forest management standards.

Global renewable plastics production

While the amount of renewable plastics produced globally is expected to grow, it was only 1% of the total global plastics production in 2021, at 2.1 million tonnes.[1]

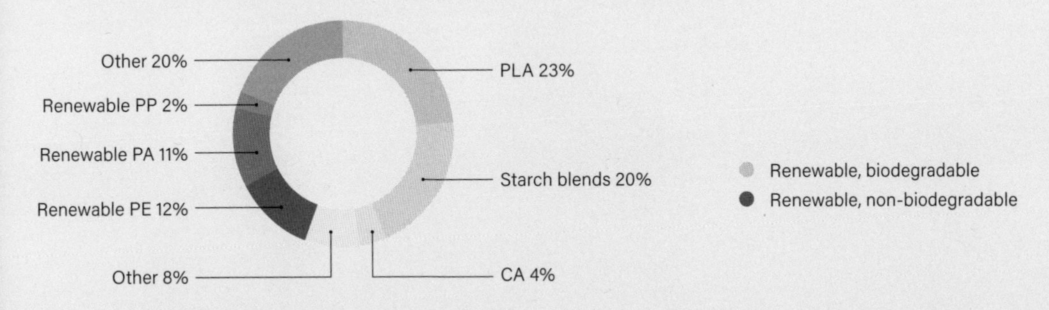

Arable land and renewable plastics

Many renewable plastics require arable land for growing raw materials, but volumes, measured in millions of hectares, are currently so small that this doesn't compete with food production. But if renewable plastics, biofuels and other materials (e.g. cotton and other fibre crops) that need arable land are going to play a larger role in the global circular economy in future, this will change.[2]

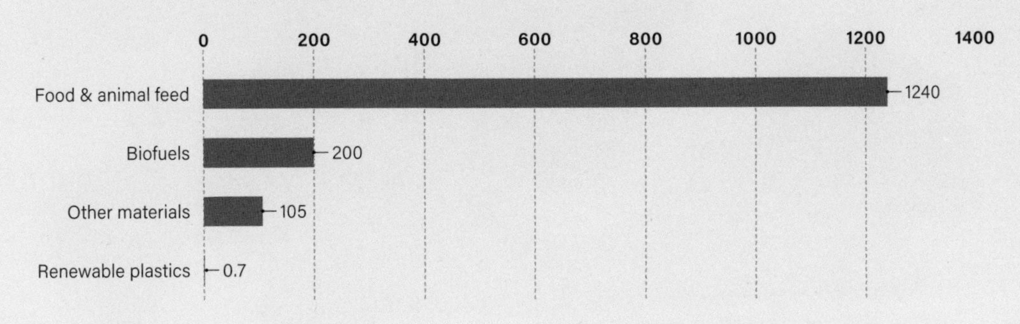

Biodegradability and renewable plastics

Renewable plastics are often assumed to be biodegradable, but this is far from the case. Of the renewable plastics in this book, only polylactide (PLA) and cellulose acetate (CA) are biodegradable. See below for an overview.

- **Home composting**
  Materials are expected to disintegrate within six months and form good compost within a year in a home compost heap. TÜV Austria's OK Compost HOME is probably the most common certification for this category. Home-compostable grades of PLA and CA are available.

- **Industrial composting**
  Materials are expected to give good-quality compost within 180 days. Common certifications include the Biodegradable Products Institute (BPI) in the US, and TÜV Austria OK Compost INDUSTRIAL in Europe. Industrially compostable grades of PLA and CA are available.

- **Biodegradable in landfill**
  This refers to materials that are expected to fully degrade within two to five years in landfill, and without adverse effects on the soil, as specified by TÜV Austria's OK Biodegradable SOIL certification. Some grades of CA are compatible.

- **Biodegradable in water**
  Materials capable of fully degrading in sea- or fresh water within six months, as specified by TÜV Austria's OK Biodegradable MARINE and OK Biodegradable WATER certification. Some grades of CA are compatible, but proceed with caution; biodegradable plastic waste in the process of degrading can cause the same problems for marine life as other plastic waste.

'Open-source renewable plastics'

An interview with Pilar Bolumburu, materials designer and researcher at Materiom, an open-source database for renewable plastics and composites that aims to speed up the transition from conventional, petrochemical-based materials. For more, visit materiom.org

With your background as an interdisciplinary designer, you've covered a lot of interesting areas in your work, such as digital fabrication and hands-on development of renewable materials. What set you off in this direction?

I first became interested in these things while studying design at university in Chile. The course I was on was very open, letting students themselves decide what modules to take and how to combine them. This meant that I was able to explore different areas of design before becoming more involved in sustainability and manufacturing, like how to produce things when there are no major industries in the local area. After university I started working at a fab lab in Santiago, getting involved in local production in a very practical way, but I also started meeting people more on the materials side of things, like Alysia Garmulewicz, who is one of the co-founders of Materiom, where I work now.

Materiom is an open-source database that can be searched based on different types of biomass, so you can find material recipes based on raw materials that are available in your local area. Combined with digital fabrication, this kind of open-source material development is really powerful. Unlike the traditional way of designing products first, then going out to source some materials based on what's available from suppliers that are often based in far-off locations, with this approach you can start by developing materials based on local resources, then design the product around that.

I was completely intrigued when I first found out about Materiom; I don't think that I've ever seen anything like it in terms of scope and vision. Could you tell me a bit about how it was established and how it works?

Alysia and another co-founder, Liz Corbin, started Materiom while they were both doing a PhD in the UK, studying the environmental and social impacts of bio-based materials. There are so many factors to take into account – natural polymers like cellulose and chitin, and other natural ingredients that can be used in materials

Agar spirulina bioplastic by Jacqueline Solis, as featured on Materiom.

production, like starches, proteins and natural fibre, can be derived from so many different sources in different places around the world. Alysia and Liz came up with the idea of creating a resource that would collect all this data on ingredients and material recipes in one place, to encourage more people to consider using renewable materials from local sources. We run a test lab, where users can submit their material recipes to receive test results for mechanical performance, environmental resistance and other properties for specific formulations. Based on the data we collect about material performance, our software uses artificial intelligence to make suggestions for how to optimize material properties, which can really help researchers and designers accelerate their R&D process to a point where these materials are ready to be used in products.

Bio-based materials, including renewable plastics, offer interesting alternatives to conventional petrochemical-based materials. But more specifically, what do you think are the key strengths and weaknesses of the materials that you're developing within the framework of Materiom?

From the beginning, Alysia and Liz were interested in both the environmental and social impact of bio-based materials. Obviously, it would have a huge impact on local communities if farmers switched from growing crops for food production to materials production. It's a complex area and we tend to use the term 'first-generation biomass' for raw materials that are grown on agricultural land and potentially in direct competition with food production. On the next level up we have 'second-generation biomass', which includes waste and by-products from first-generation biomass – things like peel, leaves, stalks and other parts of crops that are not edible. It makes sense to make use of this waste material, given the resources that have gone into growing the crops in the first place. Second-generation biomass also includes raw materials derived from the forest industry, as well as plants that don't have to be grown on agricultural land, such as bamboo. We've recently also been looking at 'third-generation biomass', including seaweed, which is a really interesting raw material for renewable plastics – completely disconnected from conventional agriculture in terms of how it's cultivated and harvested.

When you look at it this way, of course, it would be better to use only second- and third-generation biomass, but other factors also come into play, such as local conditions and culture. Here Materiom can be a big help in identifying solutions that might be a good fit locally.

In terms of material properties, I think biopolymers have huge potential in packaging, to give just one example. For sure there will be applications where recycled materials and reusable containers are a better fit, but in many cases bio-based and biodegradable materials are a much better alternative to conventional plastics, which generate a lot of waste and take a very long time to break down. I expect that in the next five to ten years we'll develop materials for which we can control the time taken to break down under different conditions, basically designing their end of life. In the longer term, we'll be able to adapt these materials to other industries, like fashion. For me, this is such a strong feature of regenerative materials – that they go back into the biological cycle when we're finished with them, as opposed to petrochemical-based materials, which remain completely separate from the cyclical processes of nature.

Materials from Materiom, left to right, from top

Row 1: marigold flowers bioplastic (Sanskriti Gupta), eggshell paste for 3D printing (Ana Otero/Coudre Studio), oyster shell composite (Xinyi Lin)

Row 2: yerba mate blown bioplastic (Ana Laura Cantera), carrageenan film (Valentina Lugae & Carolina Márquez), egg carton cornstarch card (Big Circle Studios/Zoë Powell Best)

Row 3: foamy bioplastic (Sanskriti Gupta), homemade clays (Francesca Perona & Marco Tortarolo), kombucha material textured and dyed (Josefin Åberg)

Row 4: mica bioprint (Jeremy Faludi, Corrie Van Sice & Yuan Shi), wood ashes biocomposite (LABVA/María José Besoain & Alejandro Weiss), mussel alginate composite (Carolina Pachecoa)

Renewable polypropylene (PP)

Several suppliers offer renewable PP derived from used cooking oil, waste from vegetable-oil production and other plant-based sources. The material is chemically identical to virgin petrochemical-based PP and it can be used as a drop-in alternative.

Suppliers & materials	- Borealis Bornewables™ renewable PP - LyondellBasell CirculenRenew renewable PP	
Raw material origin	Recycled cooking oil and agricultural waste, as well as virgin plant-based raw materials. Renewable PP materials are often made with a mix of renewable and petrochemical-based raw materials. Ask suppliers to provide ISCC PLUS, TÜV Austria's OK Biobased or other certification that verifies the origin and ratio of raw materials in specific renewable PP grades.	
	Borealis Bornewables™ renewable PP[3]	**Virgin petrochemical-based PP**[4]
GWP	−0.5 kg CO_2e / kg	1.6 kg CO_2e / kg
Energy use	No data	77.1 MJ / kg
Water use	No data	86.1 l / kg
Toxicity	Pure PP is considered safe for use in food-, water- and skin-contact applications. Request a material safety data sheet and ask suppliers to confirm that specific materials are REACH compliant with regards to additives.	
Circularity	Renewable PP can be recycled with other PP waste without separation. Renewable PP is not biodegradable. For an overview of recycling rates and circular design guidelines for plastics, see page 23.	
Mechanical properties	Very versatile, with low weight, good impact resistance and general toughness. Scratch resistance is rather poor, but grades with improved abrasion resistance are available.	
Environmental resistance	Good thermal resistance with a safe service temperature between 90 and 120°C (195–250°F). Good resistance to moisture and some chemicals, but UV resistance is poor.	
Forming	Straightforward to form using common thermoplastic processes, including injection moulding, blow moulding, extrusion and thermoforming.	
Finishing	Opaque and translucent renewable PP is straightforward to colour. Optically clear PP grades are also available. Most PP grades have poor scratch resistance, but abrasion-resistant grades can be specified for durable, glossy surfaces.	

Eco Conscious Edition Kettle by Philips. The outer housing is made with polypropylene derived from a variety of renewable raw material sources, including waste cooking oil and renewable naphtha made with by-products from the timber and paper industries.

PHILIPS

Renewable polyethylene (PE)

Like renewable PP (see pages 54–55), renewable PE can be derived from used cooking oil and agricultural waste, but sugar from sugarcane is also a common raw material. Renewable PE is chemically identical to virgin petrochemical-based material and can be used as a drop-in alternative.

Suppliers & materials	- Borealis Bornewables™ renewable PE - Braskem I'm green™ renewable PE - LyondellBasell CirculenRenew renewable PE	
Raw material origin	Recycled cooking oil and agricultural waste, as well as virgin plant-based raw materials such as sugar. Renewable PE materials are often made with a mix of renewable and petrochemical-based raw materials. Ask suppliers to provide ISCC PLUS, TÜV Austria's OK Biobased or other certification that verifies the origin and ratio of raw materials in specific renewable PE grades.	
	Borealis Bornewables™ renewable HDPE[5]	**Virgin petrochemical-based HDPE**[6]
GWP	−0.5 kg CO_2e / kg	1.8 kg CO_2e / kg
Energy use	No data	79.3 MJ / kg
Water use	No data	105.5 l / kg
Toxicity	Pure PE is considered safe for use in food-, water- and skin-contact applications. Request a material safety data sheet and ask suppliers to confirm that specific materials are REACH compliant with regards to additives.	
Circularity	Renewable PE can be recycled with other PE waste without separation. Renewable PE is not biodegradable. For an overview of recycling rates and circular design guidelines for plastics, see page 23.	
Mechanical properties	Available in two main varieties – high-density polyethylene (HDPE) and low-density polyethylene (LDPE). HDPE has good rigidity and tensile strength, while LDPE is more flexible and suitable for thin applications such as films.	
Environmental resistance	Good thermal and moisture resistance, as well as to some chemicals, but UV resistance is poor.	
Forming	Straightforward to form using common thermoplastic processes, including injection moulding, blow moulding, extrusion and thermoforming.	
Finishing	PE is translucent to opaque, depending on material thickness, or transparent in thin film form. It's straightforward to colour but has rather poor scratch resistance, making it unsuitable for durable, glossy surfaces.	

TePe GOOD™ toothbrush. The renewable PE material used in the handle is derived from sugarcane, while the bristles are made with renewable PA (see page 64).

TePe
GOOD
REGULAR

Polylactide (PLA)

PLA is one of the most widely used renewable plastics today, partly because of its relatively low production costs compared with many other renewable plastics, but also for its reduced environmental footprint compared with competing petrochemical-based materials – mainly polystyrene and polyethylene.

Suppliers & materials	- Total Corbion Luminy® PLA - NatureWorks Ingeo PLA
Raw material origin	Typically fermented starch from sugarcane, although other vegetables like sugar beets, cassava and corn are also used. This has an impact on food production and agriculture, so certification for sustainable sugar production that also applies to PLA materials is available from Bonsucro and others. PLA materials are typically 100% renewable.
	Total Corbion Luminy® PLA[7]
GWP	0.5 kg CO_2e / kg
Energy use	89.2 MJ / kg
Water use	60 l / kg
Toxicity	PLA is available in grades for food contact. Request a material safety data sheet and ask suppliers to confirm that materials are REACH compliant for specific PLA grades.
Circularity	While small volumes of recycled PLA are available from specialist recyclers, it's currently not widely recycled. This is likely due to the low volumes of the material in use today. Different grades of PLA are compostable at home or in industrial composting facilities. See page 49 for more about compostable and biodegradable plastics. Ask suppliers to provide industrial composting certification, such as TÜV Austria's OK Compost INDUSTRIAL or Biodegradable Products Institute (BPI).
Mechanical properties	In its basic form, PLA is a rather brittle material with low impact strength. Additives and modifiers are available to improve impact resistance, but these may impact biodegradability. Request a technical data sheet and confirm the biodegradability of specific grades.
Environmental resistance	Poor moisture and chemical resistance, in addition to low temperature resistance. There are several additives and modifiers available to improve environmental resistance performance, but again, these may impact biodegradability. Request a material safety data sheet and confirm the biodegradability of specific grades.
Forming	Compatible with common thermoplastic processes such as injection moulding, extrusion, blow moulding and thermoforming, but PLA may require significantly longer cycle times than competing petrochemical-based plastics. Grades with faster cycle times are available – request processing guidelines from suppliers.
Finishing	PLA is straightforward to colour and available in opaque and transparent grades. It has poor scratch resistance, however, making it unsuitable for durable, glossy surfaces.

Cantilever Floor Light, by Louis Filosa for Gantri. The lampshade and housing is made with a special grade of PLA developed by Gantri.

Cellulose acetate (CA)

This durable material, partially derived from wood pulp, was originally developed way back in 1864. Depending on the specific grade, CA is biodegradable in many different conditions, from home composting to landfill.

Suppliers & materials	- Eastman Tenite™ partially renewable CA - Celanese BlueRidge™ partially renewable CA - Mazzuchelli 1849 M49 68% renewable CA
Raw material origin	About 50% of the raw materials for CA consists of so-called dissolving pulp, a type of wood pulp that's also used for making cellulose-based synthetic textiles (see pages 84–85 and 102–3). The certifications that apply to sustainably sourced timber and paper – such as FSC and PEFC – also apply to CA raw materials. The remainder of the raw materials needed for CA production are normally petrochemical-based plasticizers, although Mazzuchelli and other suppliers have developed renewable plasticizers that enable a higher renewable raw material ratio.
	Cellulose acetate[8]
GWP	5.9 kg CO_2e / kg
Energy use	103.8 MJ / kg
Water use	No data
Toxicity	While CA production involves harsh chemicals, finished CA material is considered safe for food and skin contact. The material is often used in fashion accessories such as eyewear, where it's in close contact with the skin for prolonged periods. Request a material safety data sheet and confirm the suitability of specific grades for food- and skin-contact applications.
Circularity	Recycled CA is available from specialist recyclers, such as Mazzuchelli's BeCycle™ programme, but it's currently not widely recycled. CA is generally biodegradable, with some grades suitable for home and/or industrial composting, and in some cases it may safely degrade in landfill. For more about compostable and biodegradable plastics, see page 49. Confirm the biodegradability of specific grades with suppliers, as well as any applicable certification.
Mechanical properties	Tough and durable, with excellent impact strength. While the material surface is quite soft, light scratches will usually disappear if polished gently.
Environmental resistance	Good moisture and thermal resistance, but only moderate chemical resistance.
Forming	Supplied in semi-formed sheets for further machining, thermoforming or other processing, as well as grades for injection moulding and extrusion.
Finishing	Available in opaque and transparent grades. The highly decorative translucent CA materials often seen in eyewear and other fashion accessories are typically made from cast CA sheets, where different colours can be mixed to great effect. CA is a good fit for durable, polished surfaces.

Prism Tortoise Watch by AARK. The watch's housing is made with cellulose acetate derived from wood pulp.

Renewable polycarbonate (PC)

Renewable PC offers the same high performance and optical clarity as its petrochemical-based equivalent, but with a significantly lower carbon footprint.

Suppliers & materials	- SABIC TRUCIRCLE™ LEXAN™ renewable PC - Covestro Makrolon® RE renewable PC	
Raw material origin	Recycled cooking oil and agricultural waste, as well as virgin plant-based raw materials. However, renewable PC materials are typically made with a mix of renewable and petrochemical-based raw materials. Ask suppliers to provide ISCC PLUS, TÜV Austria's OK Biobased or other certification that verifies the origin and ratio of raw materials in specific renewable PC grades.	
	SABIC TRUCIRCLE™ LEXAN™ renewable PC[9]	**Virgin petrochemical-based PC**[10]
GWP	1.3 kg CO_2e / kg	3.4 kg CO_2e / kg
Energy use	No data	99 MJ / kg
Water use	No data	1,535 l / kg
Toxicity	Concerns have been raised about Bisphenol A (BPA), a key chemical ingredient in PC, because of possible links to hormone disruption in humans and animals, so PC should be avoided for food- and water-contact applications.	
Circularity	Renewable PC can be recycled with conventional petrochemical-based PC without separation. It's not biodegradable. For an overview of recycling rates and circular design guidelines for plastics, see page 23.	
Mechanical properties	Lightweight, tough and impact resistant, with good abrasion resistance.	
Environmental resistance	Good UV and thermal resistance, as well as resistance to certain chemicals.	
Forming	Typically supplied in pellet form for injection moulding, extrusion and other common thermoplastic process, as well as in semi-formed sheets for further forming through machining and thermoforming.	
Finishing	Very good optical clarity and easy to colour using pigments or dyes. PC has good scratch resistance, making it suitable for durable, glossy surfaces.	

EVBox Livo home charging station for electric vehicles. The exterior housing is made with Makrolon® RE renewable PC from Covestro.

EVBOX

Renewable polyamide (PA)

Renewable PA (also called nylon) can significantly reduce the sizable environmental footprint of conventional petrochemical-based PA, without compromising the performance of this tough and durable material.

Suppliers & materials	- DSM EcoPaXX® renewable PA410 - Arkema Rilsan® renewable PA11 - Evonik VESTAMID® Terra PA1010	
Raw material origin	Typically oil from castor beans, a plant that's not used in food production and can be grown on arid, non-agricultural land. Renewable PA is often made with a mix of renewable and petrochemical-based raw materials. Ask suppliers to provide ISCC PLUS, TÜV Austria's OK Biobased or other certification that verifies the origin and ratio of raw materials in specific renewable PA grades.	
	DSM EcoPaXX® renewable PA410[11]	**Virgin petrochemical-based PA6**[12]
GWP	1.9 kg CO_2e / kg	6.7 kg CO_2e / kg
Energy use	No data	128.8 MJ / kg
Water use	No data	1,647 l / kg
Toxicity	PA is often used in food-, water- and skin-contact applications, but it will give off toxic fumes if heated above 300°C (570°F).	
Circularity	PA recycling is somewhat complex due to the many different types of PA in common use. While it's possible to recycle all types of PA without separation, the performance of the resulting material is unpredictable. Ideally, each type of PA should be sorted and recycled separately. For example, Arkema runs its own recycling programme for its portfolio of PA11-based materials. Renewable PA is not biodegradable. For an overview of recycling rates and circular design guidelines for plastics, see page 23.	
Mechanical properties	Excellent impact strength and stiffness, combined with good abrasion and wear resistance.	
Environmental resistance	Good chemical, UV, moisture and thermal resistance.	
Forming	Straightforward to form using common thermoplastic processes, including injection moulding, extrusion and thermoforming.	
Finishing	Available in opaque and clear grades that are easy to colour. PA has good abrasion resistance, making it suitable for durable, glossy surfaces.	

Swatch® REDVREMYA watch. The housing is made with a plastic material derived from castor beans, a non-food crop that can grow in desert-like conditions.

swatch

Renewable liquid epoxy resin (LER)

Petrochemical-based LER is one of the most widely used thermosetting plastics, but it comes with a sizable environmental footprint. Renewable LER offers significant reductions without compromising performance.

Suppliers & materials	- Spolchemie ENVIPOXY® renewable LER - Sicomin GreenPoxy renewable LER - Entropy Resins renewable LER	
Raw material origin	Raw materials for renewable LER production are typically derived from soybeans, palm oil, or tallow from livestock. However, renewable LER materials are typically made with a mix of renewable and petrochemical-based raw materials. Ask suppliers to provide ISCC PLUS, TÜV Austria's OK Biobased or other certification that verifies the origin and ratio of raw materials in specific renewable LER grades.	
	Spolchemie ENVIPOXY® renewable LER[13]	**Petrochemical-based LER**[14]
GWP	1.3 kg CO_2e / kg	4.5 kg CO_2e / kg
Energy use	89.3 MJ / kg	94.1 MJ / kg
Water use	12,500 l / kg	3,180 l / kg
Toxicity	In its uncured liquid form, LER gives off toxic fumes and causes skin irritation. However, once hardened it's considered non-toxic and several epoxy materials are approved for food contact. Request a material safety data sheet and confirm the suitability of specific grades for food- and skin-contact applications.	
Circularity	LER is currently not widely recycled, mainly because it's a thermoset material and cannot easily be mechanically recycled like thermoplastics can. This may change in the future, as chemical recycled processes become more established.	
Mechanical properties	Exceptional tensile and flexural strength, as well as good abrasion resistance, but rather brittle and prone to breaking on impact.	
Environmental resistance	Good resistance to UV radiation and heat, as well as moisture and chemicals.	
Forming	Common LER forming processes include casting, compression moulding, **resin transfer moulding** and **reaction injection moulding**.	
Finishing	LER is available in opaque and transparent grades that are straightforward to colour. It has good scratch resistance, making it suitable for durable, glossy surfaces.	

Aura Light, designed by Sabine Marcelis for Established & Sons. The lampshade is made with epoxy resin partially derived from vegetable glycerol.

Renewable thermoplastic elastomers (TPEs)

The advantage of TPEs over thermoset elastomers such as natural rubber and silicone is that they can be formed using common thermoplastic processes such as injection moulding and extrusion. Renewable TPEs offer the same performance and forming potential as petrochemical-based TPE, with a reduced environmental footprint.

Suppliers & materials	- TRINSEO™ APILON™ 52 BIO renewable TPU (thermoplastic polyurethane) - DSM Arnitel® Eco renewable TPC (thermoplastic copolyester) - Arkema Pebax® renewable TPA (thermoplastic polyether block amide) - HEXPOL Dryflex® Green renewable TPS (thermoplastic styrenic block copolymer)	
Raw material origin	A wide range of plants – from food crops such as sugarcane and sugar beets to vegetable oil, and non-food sources such as cellulose and castor beans. The renewable content in renewable TPE materials usually varies between 15 and 50%, and is often less in softer grades. Ask suppliers to provide ISCC PLUS or other certification that confirms the origin and ratio of raw materials used in specific grades.	
	TRINSEO™ APILON™ 52 BIO renewable TPU[15]	**Virgin petrochemical-based TPU**[16]
GWP	2.3 kg CO_2e / kg	4.7 kg CO_2e / kg
Energy use	No data	50 MJ / kg
Water use	40 l / kg	59 l / kg
Toxicity	TPEs are common in food-, water- and skin-contact applications, but the chemical building blocks of different TPE materials vary considerably, so always request a material safety data sheet and confirm the suitability of specific grades for specific applications.	
Circularity	TPEs are currently not widely recycled, possibly because of relatively low volumes compared with many other plastics, but also because they're often used in combination with other materials, so to maximize the likelihood of recycling, try to make TPE parts easily separable, or make the entire product from TPE if possible. Sometimes specific TPEs can be recycled with other materials without separation – e.g., Arkema's Pebax® TPA elastomers can be recycled with their Rilsan® PA11 materials. Sometimes specific TPEs can be recycled with other materials without separation. For an overview, see page 23.	
Mechanical properties	Generally tough, with good tear and abrasion resistance. All TPEs can be stretched and allowed to return to their original shape without permanent deformation. Suppliers typically offer several grades, with different flexibility and softness properties.	
Environmental resistance	Generally good temperature resistance, but UV, moisture and chemical resistance varies. Request a technical data sheet and confirm the suitability of specific TPE grades for specific applications.	
Forming	Unlike thermoset elastomers, TPEs can be formed using common thermoforming processes such as injection moulding, extrusion and thermoforming.	
Finishing	Renewable TPEs are available in opaque and translucent grades and are typically easy to colour using pigments. They are warm and tactile, with good potential for in-mould textures.	

Link-It pen by Schneider. The grip is made with a renewable TPE material derived from agricultural products and waste.

Schneider
Link-It

Natural rubber

Natural rubber, also known as latex, is a renewable thermoset elastomer with a history going back thousands of years. It's useful in its raw state, tapped directly from the rubber tree, but natural rubber is more commonly vulcanized – a process for hardening the material and making it more durable.

Suppliers & materials	- Yulex FSC-certified natural rubber - Weber & Schaer PEFC- and FSC-certified natural rubber
Raw material origin	Most natural rubber is derived from rubber-tree plantations, but several research projects have been launched to extract natural rubber from other sources, including dandelions and guayule, a shrub that grows in arid conditions. Both FSC and PEFC certification is available for sustainable natural rubber production.
	Natural rubber (vulcanized)[17]
GWP	2.5 kg CO_2e / kg
Energy use	42 MJ / kg
Water use	10 l / kg
Toxicity	Natural rubber can be used in food- and water-contact applications, but both the material itself and some of the additives used during vulcanization can cause allergic reactions. Request a material safety data sheet and confirm the suitability of specific natural rubber grades for specific applications.
Circularity	Natural rubber is fairly widely recycled, mainly in the form of used tyres; about half of all used tyres in Europe are recycled currently.[18] Natural rubber is often used in combination with other materials, so to maximize the likelihood of recycling, try to make natural rubber parts easily separable from other materials, or make the entire product from natural rubber. While natural rubber is biodegradable, technically speaking, it may take up to 50 years to degrade, depending on environmental conditions, as well as the type of rubber and additives used.
Mechanical properties	Tough and flexible, with excellent abrasion and tear resistance.
Environmental resistance	Good resistance to water and certain chemicals, but natural rubber doesn't perform as well in terms of resisting UV radiation and temperatures above 80°C (175°F).
Forming	Common forming processes include casting, dip moulding and compression moulding.
Finishing	Natural rubber is opaque and pigments can be added for colour. It's a warm and tactile material with good potential for surface textures.

Nieuwland 2e Yulex Long Sleeve Swimsuit by Finisterre. Yulex is a hardwearing, yet soft and flexible material made with natural rubber from FSC-certified rubber plantations.

2 Textiles

Although used in much larger volumes in other industries such as fashion, textiles also play an important role in product design – used for furniture and interiors, consumer electronics and automotive interiors – and as a result, have a sizable environmental footprint in this field. They form a complex family of materials, with natural textiles dating back thousands of years, alongside the relatively recent introduction of synthetic textiles, intrinsically linked to the development of plastic materials.

There are broadly two types of textiles: woven textiles include any fabric that's woven or knitted, while non-woven textiles include leather, synthetic membranes and films, and bonded-fibre textiles like felt. The diagram opposite outlines the textiles in this book – it doesn't come close to mapping the full array of textile materials in common use today. Added to this complexity is the difficulty of assessing environmental impact based on raw material alone. Fibres and filaments typically require a significant amount of further processing, including spinning fibres into yarn or extruding synthetic raw materials into filament. Weaving and knitting with different yarn thicknesses has a major impact on the environmental footprint of the finished textile. Each step of the process also typically relies on a vast array of chemicals – from synthetic textiles that are made entirely with chemicals, to the fertilizers and pesticides that are widely used for growing raw materials like cotton, to spinning oils and sizing chemicals for the pre-treatment of yarns and textiles. Finishing processes like dyeing and printing add yet another layer, with implications for the toxicity, water footprint and durability of textiles.

There are several excellent resources to help designers make sense of the complex world of textiles and sustainability. One is the *Preferred Fiber and Materials Market Report* produced by Textile Exchange, a non-profit organization promoting sustainable textiles. There are also several certifications aimed at reducing the environmental and societal impact of textiles. OEKO-TEX© and the Global Organic Textile Standard (GOTS) track the overall industry; others focus on specific raw materials and are listed with the relevant materials in this book. Some of these organizations also offer supplier databases to help with sourcing sustainable textiles, such as OEKO-TEX©'s Buying Guide, the GOTS Certified Suppliers Database and Textile Exchange's supplier database.

Textile fibres, filaments and materials

The textile world consists of a wide array of fibres, filaments and materials. This family tree gives an overview of the textile materials featured in this book and where they belong in terms of type.[1]

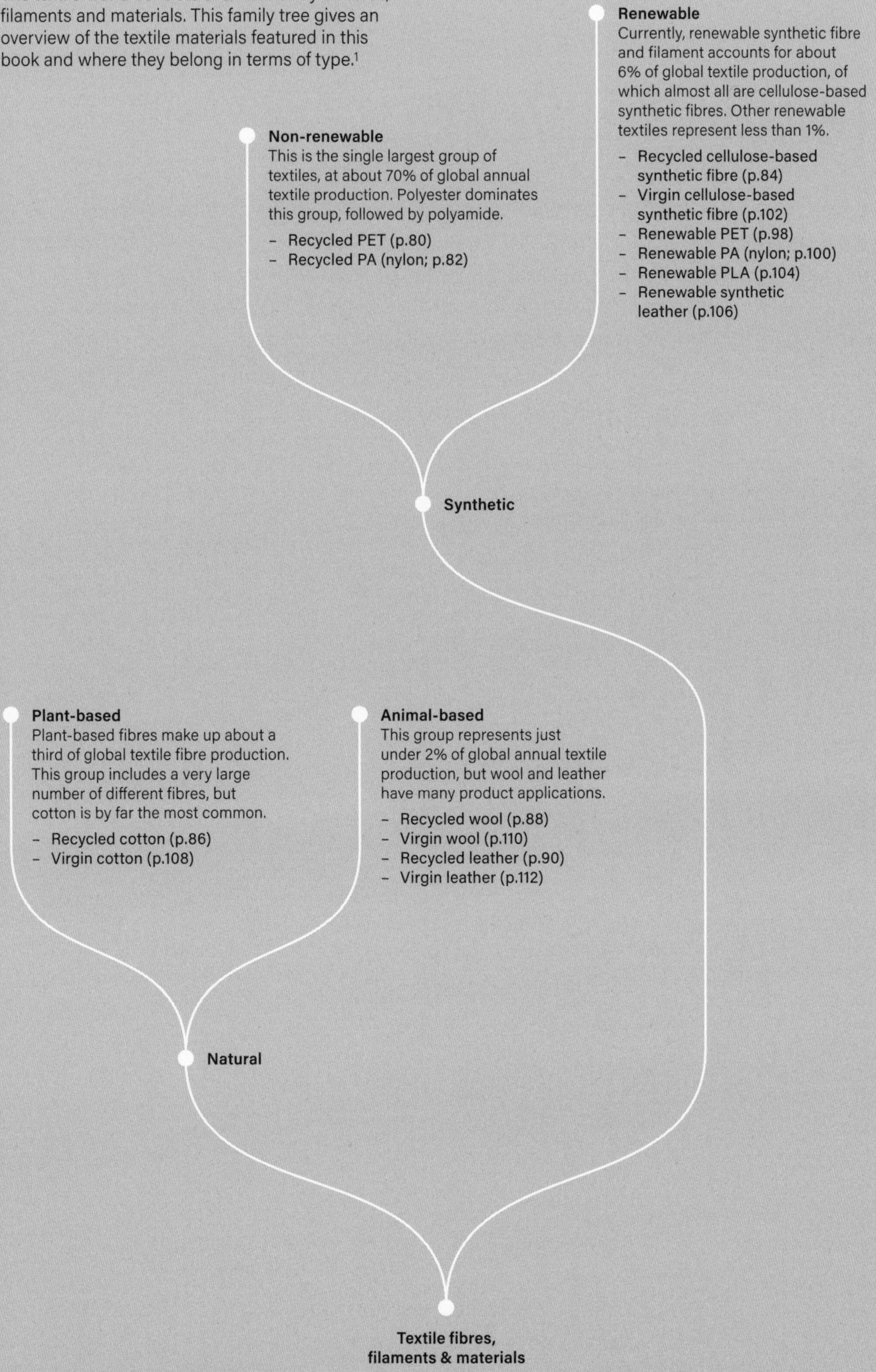

Renewable

Currently, renewable synthetic fibre and filament accounts for about 6% of global textile production, of which almost all are cellulose-based synthetic fibres. Other renewable textiles represent less than 1%.

- Recycled cellulose-based synthetic fibre (p.84)
- Virgin cellulose-based synthetic fibre (p.102)
- Renewable PET (p.98)
- Renewable PA (nylon; p.100)
- Renewable PLA (p.104)
- Renewable synthetic leather (p.106)

Non-renewable

This is the single largest group of textiles, at about 70% of global annual textile production. Polyester dominates this group, followed by polyamide.

- Recycled PET (p.80)
- Recycled PA (nylon; p.82)

Synthetic

Plant-based

Plant-based fibres make up about a third of global textile fibre production. This group includes a very large number of different fibres, but cotton is by far the most common.

- Recycled cotton (p.86)
- Virgin cotton (p.108)

Animal-based

This group represents just under 2% of global annual textile production, but wool and leather have many product applications.

- Recycled wool (p.88)
- Virgin wool (p.110)
- Recycled leather (p.90)
- Virgin leather (p.112)

Natural

Textile fibres, filaments & materials

Textile recycling

Textile recycling can be relatively straightforward – a large piece of discarded fabric can, for example, be cut into smaller pieces and repurposed into new products. From a circularity point of view, this is an ideal scenario as it uses very little energy, but it may not be a good fit in cases where reliability and consistency in terms of availability and quality is important.

In these cases, recycled yarns and filaments can be a better option, enabling recycled fabrics to be made to spec. It's worth pointing out, however, that while some suppliers offer recycled textiles made from actual textile waste, recycled textile fibres are much more commonly made using waste materials from other industries altogether, such as discarded PET bottles and polyamide (PA, or nylon) fishing nets. The main reason for this is that it's fairly difficult to re-spin fibre from mechanically shredded textile waste into recycled yarn using the processes that are available today.

Another aspect to bear in mind with textiles in the context of circularity is that many textiles consist of blends of different textile fibres, such as cotton and elastane (aka spandex) in stretch fabrics, and cotton and polyester (PET) for reducing static and 'pilling' – the little 'pills' of entangled fibres that can form on the surface of textiles due to wear. Currently, only a few specialist recyclers are capable of recycling mixed-fibre textiles, so these materials should be avoided and any desired properties – whether anti-static, stretch or something else altogether – should ideally be achieved with a single textile material that's more likely to be recycled at the end of the life of the product.

In addition to conventional mechanical recycling, chemical recycling processes for textiles are being developed, with a handful of commercially available materials, such as Ambercycle™ Cycora chemically recycled PET textiles (see page 80) and Lenzing REFIBRA™ PIR Lyocell, partially made with chemically recycled cotton (see page 84). Chemical recycling offers many advantages over mechanical recycling, such as the ability to recycle mixed textile waste in some cases, as well as removing any additives such as dyes, inks and other finishes, meaning that the recycled fibre can be finished in the same way as virgin material. The downside is that chemical recycling processes are typically considerably more energy-intensive than mechanical recycling processes, at least for now.

Global textile recycling rates

Textiles are currently not widely recycled and many 'recycled' textiles are not made with recycled textiles at all – most recycled polyester is made with waste PET bottles, while recycled PA often uses waste from other industries, from items such as fishing nets and carpets. Currently, most actual textile recycling includes cotton, mostly going into recycled synthetic cellulose-based textiles, and recycled wool, covering about 6% of the global demand for wool textiles.[1]

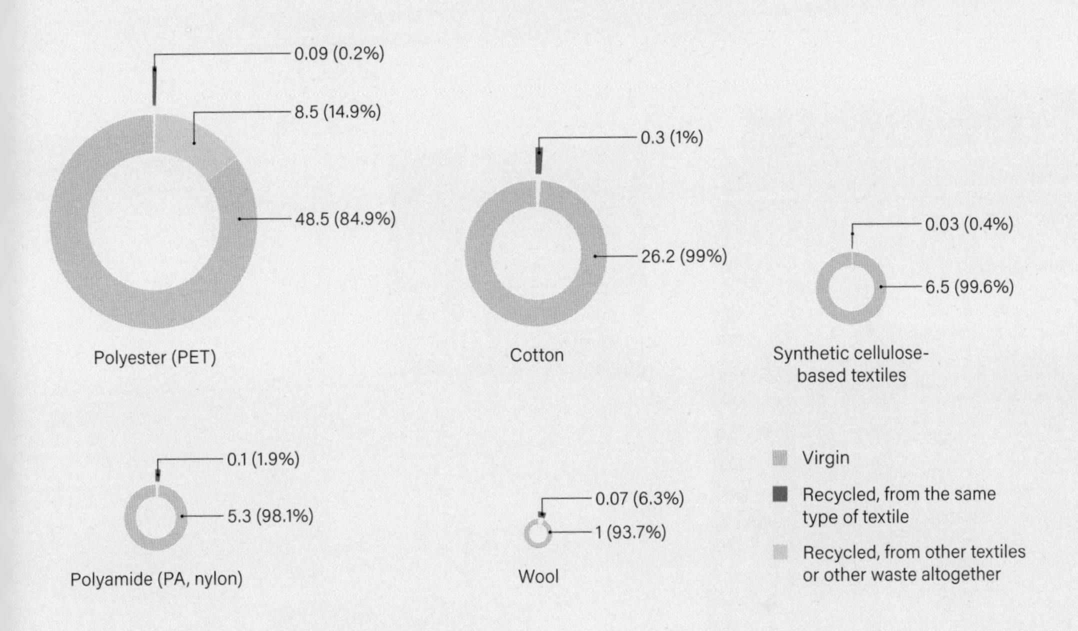

Textile waste streams

Typically, individual fibre and filament types in mixed-fibre textiles need to be extracted and separated before recycling, so it can be less complicated to produce recycled textiles from other waste, such as plastic packaging.

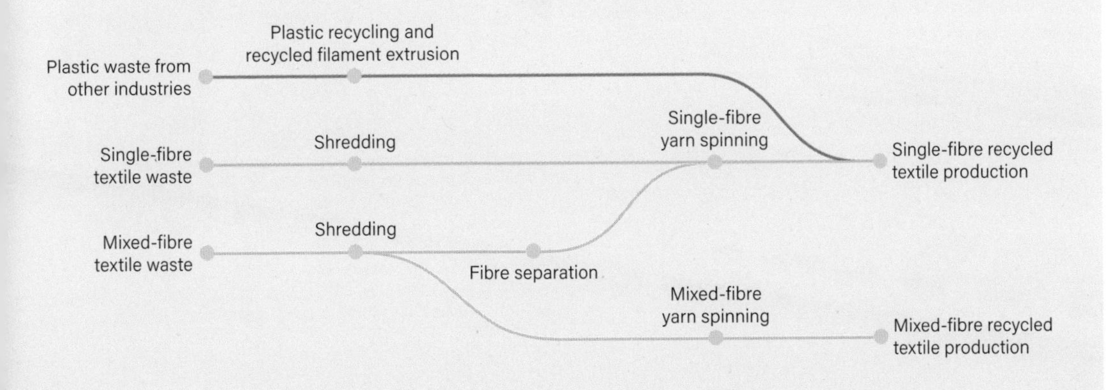

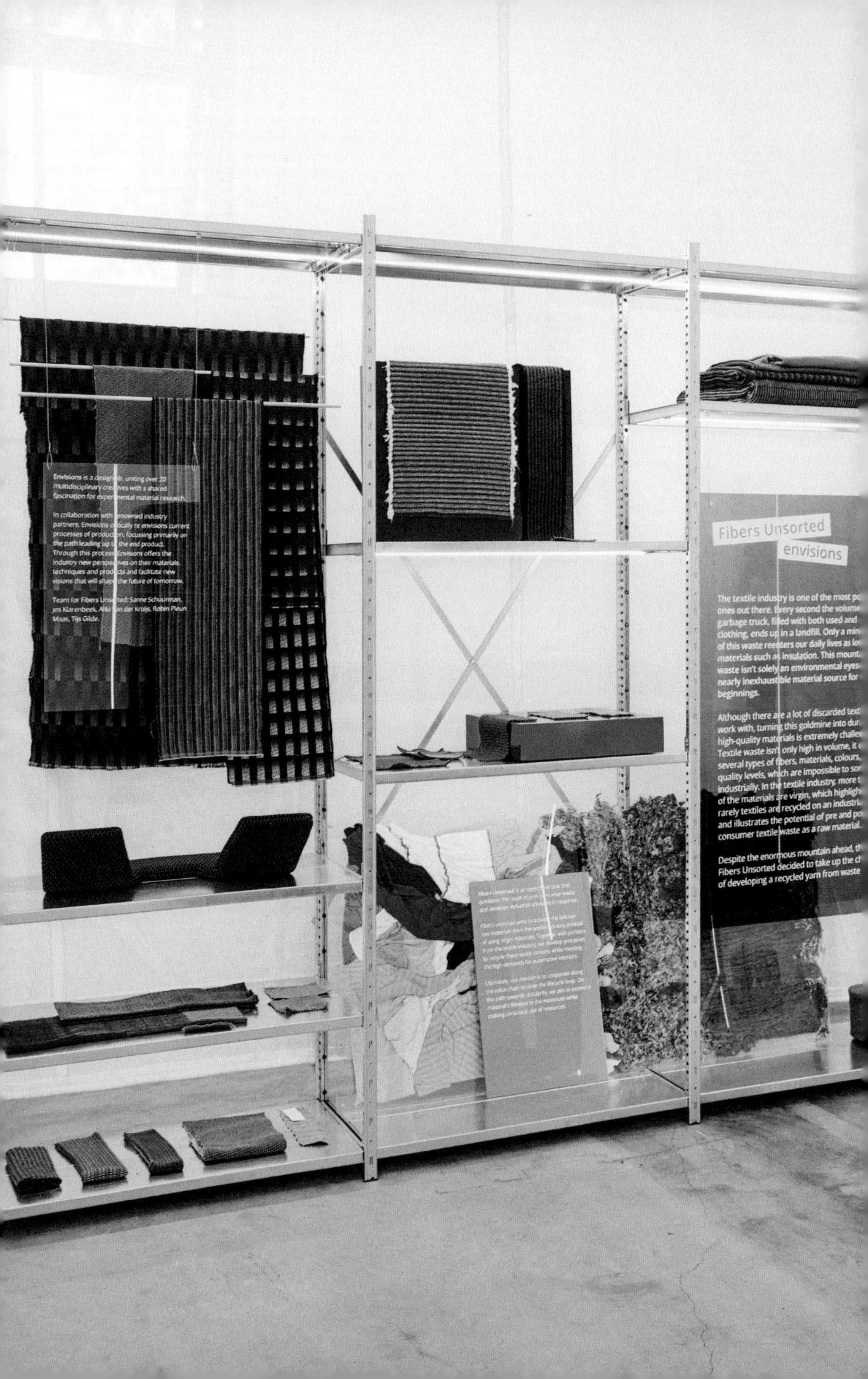
Fibers Unsorted
envisions
processes of production, focussing primarily on
Maas, Tijs Gilde.

'The volume of valuable textiles that is thrown away is shocking to me.'

An interview with Michael Wolf, textile design engineer and designer, and inventor of the Fibers Unsorted textile recycling process, which makes it possible to re-spin yarn from mixed textile waste. A selection of textiles designed in collaboration with the Dutch design studio Envisions was shown at Dutch Design Week in 2021. Find out more at fibersunsorted.com

The very wide range of raw materials in common use in the textile industry can make even an outsider like me appreciate the difficulties with recycling textiles, and especially mixed-fibre textiles. Nevertheless, you have developed a process for re-spinning yarns from mixed-fibre textile waste. Could you tell me a little about the development of the Fibers Unsorted process?

The starting point was a visit to a textile recycling facility in the Netherlands that I took when I was a student. This was the first time that I realized just how much mixed-fibre textile waste is burnt or sent to landfill. Even at a small facility like the one I visited, the volumes of valuable textiles that were being thrown away were shocking to me, and what little of this waste was recycled was effectively downcycled into insulation and other materials completely unlike the textiles that went into the recycling process.

So when I returned to my university, I wanted to start a project exploring different processes for recycling this waste and I asked my professor if I could have access to the machines in the workshop to carry out some trials. When I first started bringing in waste textiles from the recycling facility, the technicians were sceptical, saying things like 'Why do you want to do this?' and 'This material is absolute rubbish' – and in a way they were right. But after a few months, when I was able to present the first results, they could see that maybe there was something worth exploring.

Fibers Unsorted exhibition at the Salone del Mobile, 2022.

Around the same time I'd discovered through my research that re-spinning yarns with mixed-fibre waste had not been uncommon in the past, so I knew that it would be possible if I could just modify the standard spinning machines, which are optimized for virgin material, to allow for using waste materials instead. So I started contacting machine manufacturers to find out how these technical problems could be solved.

I have to be honest and say that this was a slow and sometimes frustrating process, but along the way I could sense that projects like this are increasingly being backed by new legislation regarding the mandatory use of recycled materials in industries that are really important for textile manufacturers, such as the automotive industry. So there was growing interest, and I managed to secure funding and the support of machine manufacturers to develop the yarns further.

The first yarns I made were quite coarse and maybe not the most durable, but I was able to use them as proof of concept. Since then, I have developed thinner and more robust yarns that are more in line with the functional requirements for the automotive industry and other applications like furniture and interiors. I'm also experimenting with different spinning machines, such as those that use ring spinning, which allows for stronger yarns and higher production speeds.

When I saw the Fibers Unsorted exhibition in 2021, I was struck by how beautiful the textiles were. I know that you worked with Sanne Schuurman and the Envisions team on the development. How did this collaboration come about?

Because I'm working with mixed-textile waste, the recycled yarn and textiles that I can make have a greyish beige colour, which is not the most exciting. So I knew I had to make something that would be more appealing to designers and brands. I have followed Envisions since the start and I knew that they were the perfect partners to explore the aesthetic boundaries of the process. In the end, they came up with this clever method of using the recycled yarn in the weft direction of the woven textiles, and using a contrasting, virgin yarn in the warp direction. This approach opens up a much wider range of expressions, and also improves the overall performance and reliability of the textiles, making them suitable for the demanding requirements of the automotive industry, while also maintaining the highest possible amount of recycled content.

Fibers Unsorted exhibition at Dutch Design Week 2021.

Recycled polyester (PET) textiles

The vast majority of recycled PET textiles are made with waste PET bottles - very few actual polyester textiles are recycled. While this is of course not ideal in terms of circularity, PET bottle-based textiles do offer a reduced environmental footprint compared with virgin PET textiles, without loss of quality.

Suppliers & materials	- Teijin ECOPET® PCR PET textile - Unifi REPREVE® PCR PET textile - Ambercycle™ Cycora® PCR PET textile using chemical recycling	
Raw material origin	Mainly recycled PET bottles, but some recyclers, such as Ambercycle™, are capable of recycling polyester textiles using chemical recycling processes. Ask suppliers to provide ISCC PLUS, SCS Recycled Content or other certification that verifies the ratio of recycled content.	
	Teijin ECOPET® PCR PET textile[2]	**Virgin petrochemical-based PET textile**[3]
GWP	1 kg CO_2e / kg filament	2.7 kg CO_2e / kg filament
Energy use	11.9 MJ / kg filament	78.4 MJ / kg filament
Water use	No data	No data
GWP	3.9 kg CO_2e / kg woven textile (500 dtex)	5.7 kg CO_2e / kg woven textile (500 dtex)
Toxicity	Washing PET textiles is believed to be a major cause of the spread of plastic microfibres, but other than this, pure PET textiles are considered non-toxic. Certain ingredients used in textile finishing processes have been linked to toxicity. Request a material safety data sheet for dyes, colourants and other finishes, and confirm REACH compliance and other relevant certification with suppliers.	
Circularity	Recycled PET textiles accounted for around 15% of the total 109 million tonnes of PET textiles produced globally in 2020. Some 99% of recycled PET textiles were made with waste PET bottles.[4] Avoid combinations with other types of textile fibre to increase the likelihood of recycling on disposal. For an overview of recycling rates and circular design guidelines for textiles, see page 75.	
Mechanical properties	Strong and durable, yet lightweight and flexible. Weaving tends to result in stronger, more durable textiles, while knitted textiles offer more flexibility. PET felt gives good cushioning.	
Environmental resistance	Water resistance is good, although PET textiles tend to absorb grease, making stain removal difficult. Unlike natural textiles, PET textiles are not breathable.	
Forming	Weaving and knitting, as well as felting, artificial down and other non-woven processes. The material can be further formed using common textile processes such as cutting, sewing and ultrasonic welding.	
Finishing	The use of clear PET bottles makes it almost as straightforward to colour recycled textiles as virgin material. Chemically recycled textiles can be finished like virgin material. Compatible with many textile finishing processes, including printing, embossing and debossing, transfers and decorative lamination, although the latter should be PET-based so as not to impact circularity.	

Nest Mini speaker by Google. The cover is made with a recycled polyester textile made with waste PET bottles.

Recycled polyamide (PA) textiles

Similar to recycled PET textiles (pages 80–81), the waste materials used in recycled PA (or nylon) textiles typically come from other industries. The environmental impact of virgin PA is relatively large, and recycled PA textiles offer considerable reductions.

Suppliers & materials	- Aquafil ECONYL® PCR PA6 textile - Chainlon GREENLON® PCR PA6.6 textile - Unifi REPREVE® PCR PA6.6 textile	
Raw material origin	Typically recovered fishing nets, carpets, moulded PA parts and textile scraps. Ask suppliers to provide ISCC PLUS, SCS Recycled Content or any other certification that verifies the ratio of recycled content.	
	Aquafil ECONYL® PCR PA6 textile[5]	**Virgin petrochemical-based PA6 textile**[6]
GWP	3.2 kg CO_2e / kg filament	9.6 kg CO_2e / kg filament
Energy use	No data	148 MJ / kg filament
Water use	11 l / kg filament	No data
GWP	4.7 kg CO_2e / kg woven textile (500 dtex)	12.6 kg CO_2e / kg woven textile (500 dtex)
Toxicity	As with PET textiles, washing PA textiles is believed to be a source of microfibre proliferation, although to a lesser degree. Pure PA textiles are considered non-toxic, but certain ingredients used in textile finishing processes have been linked to toxicity. Request a material safety data sheet for dyes, colourants and other finishes, and confirm REACH compliance and other relevant certification with suppliers.	
Circularity	Recycled PA textiles accounted for around 1.9% of the total 5.4 million tonnes of PA textiles produced globally in 2020.[7] Avoid combinations with other types of textile fibre to increase the likelihood of recycling on disposal. For an overview of recycling rates and circular design guidelines for textiles, see page 75.	
Mechanical properties	Woven PA textiles can be extremely tough and hardwearing; knitted PA textiles are generally more elastic and less durable.	
Environmental resistance	Generally sensitive to heat, with a melting point of around 100°C (210°F). PA6 and 6.6 textiles readily absorb moisture, so odours are potentially retained inside the textiles. Other types of PA textiles have better moisture resistance.	
Forming	Weaving and knitting, as well as felting and other non-woven processes. The material can be further formed using common textile processes such as cutting, sewing and ultrasonic welding.	
Finishing	The colour of recovered PA waste will impact colourability, although careful sorting by colour, overdyeing and mixing with virgin material can allow suppliers to offer a wider range of colour options. Compatible with common textile finishing processes, including printing, embossing and debossing, transfers and decorative lamination, although the latter should be PA-based so as not to impact circularity.	

Fairphone charging cable. The protective woven textile sheathing is made with recycled polyamide.

Recycled cellulose-based synthetic textiles

Perhaps a little confusingly, most commercially available recycled cellulose-based synthetic textiles are made from cotton textile waste. This cotton waste is converted into so-called dissolving pulp using a chemical recycling process, and can then be further processed into viscose, lyocell and other cellulose-based textile fibre.

Suppliers & materials	- Renewcell Circulose® viscose fibre made with PCR cotton waste - Sateri FINEX™ PIR/PCR viscose fibre - Birla Cellulose™ LIVA REVIVA PIR viscose fibre	
Raw material origin	Depending on the specific recycling process, PIR and PCR cellulose-based textile waste can be used. Ask suppliers to provide relevant certification that verifies the ratio and type of recycled content.	
	Viscose fibre made with recycled pulp[8]	**Virgin viscose fibre**[9]
GWP	0.6 kg CO_2e / kg fibre	1.5 kg CO_2e / kg fibre
Energy use	No data	20.6 MJ / kg fibre
Water use	No data	No data
GWP	2.3 kg CO_2e / kg woven textile (500 dtex)	3.2 kg CO_2e / kg woven textile (500 dtex)
Toxicity	There are four basic types of cellulose-based synthetic textile fibres - viscose (or rayon), lyocell, modal and cupro. Various chemicals are used in the fibre manufacturing process, including some that are hazardous. The use of these chemicals can be very harmful to factory workers and the local environment unless properly managed. Request a material safety data sheet for specific fibres, dyes, colourants and other finishes, and confirm REACH compliance and relevant certification with suppliers.	
Circularity	Recycled textiles accounted for around 0.4% of the total 6.5 million tonnes of cellulose-based synthetic textiles produced globally in 2020.[10] This is expected to grow as chemical recycling processes become more established. Avoid combinations with other types of textile fibre to increase the likelihood of recycling on disposal. Some cellulose-based synthetic textiles are biodegradable; ask suppliers for confirmation and relevant certification. For an overview of recycling rates and circular design guidelines for textiles, see page 75.	
Mechanical properties	Typically soft and rather delicate, although durability and toughness depend on the yarn count and textile construction. These fibres can also be used for very robust technical textiles.	
Environmental resistance	Breathable and absorbent. While cellulose-based synthetic textiles have poor resistance to hot water and UV radiation, some suppliers offer specially treated textiles with improved UV resistance.	
Forming	Weaving and knitting, as well as felting and other non-woven processes. The material can be further formed using common textile processes such as cutting and sewing.	
Finishing	Textiles from chemically recycled textiles can typically be finished like virgin material using many textile finishing processes, including printing and dyeing.	

T-shirt by Swole Panda. The knitted textile in the T-shirt uses TENCEL™ Lyocell x REFIBRA™ lyocell fibre from Lenzing AG, made with a mix of PCR cotton waste and virgin lyocell fibre.

Recycled cotton textiles

Currently, cotton recycling largely uses post-industrial waste, although some recyclers are also capable of processing post-consumer cotton textiles. Depending on the quality of the recycled yarn, recycled content varies considerably, with suppliers offering textiles with 15 to 100 per cent recycled fibre content.

Suppliers & materials	- Recover™ RPure PIR and PCR cotton - Circular Systems Texloop™ RCOT™ Primo PIR cotton	
Raw material origin	Mostly PIR waste, with some suppliers able to process PCR waste as well. Ask suppliers to provide ISCC PLUS, SCS Recycled Content or other certification that verifies the ratio and type of recycled content.	
	Recover™ RPure recycled cotton[11]	**Virgin cotton**[12]
GWP	0.2 kg CO_2e / kg fibre	1.8 kg CO_2e / kg fibre
Energy use	1.3 MJ / kg fibre	15 MJ / kg fibre
Water use	No data	2,100 l / kg fibre
GWP	1.7 kg CO_2e / kg woven textile (500 dtex)	4.3 kg CO_2e / kg woven textile (500 dtex)
Toxicity	Pure cotton textiles are considered non-toxic, but certain ingredients used in textile finishing processes have been linked to toxicity. Request a material safety data sheet for dyes, colourants and other finishes, and confirm REACH compliance and other relevant certification with suppliers.	
Circularity	Recycled cotton textiles only accounted for around 0.96% of the total 26.2 million tonnes of cotton textiles produced globally in 2020.[13] This is expected to grow as emerging recycling processes become more established. Avoid combinations with other types of textile fibre to increase the likelihood of recycling on disposal. Pure cotton textiles are compostable at home or industrially. For an overview of recycling rates and circular design guidelines for textiles, see page 75.	
Mechanical properties	Soft and rather delicate, although durability and toughness depend on the thickness of yarn and the process used – woven textiles tend to be stronger than knits, for example.	
Environmental resistance	Poor UV resistance, so colours will typically fade over time. Good breathability and thermal resistance, although cotton textiles may shrink when washed at high temperatures.	
Forming	Weaving and knitting, as well as felting and other non-woven processes. The material can be further formed using common textile processes such as cutting and sewing.	
Finishing	The colour of recovered cotton textile waste will impact colourability, although careful sorting by colour, overdyeing and mixing with virgin material can allow suppliers to offer a wider range of colour options. Compatible with many textile finishing processes, including printing and dyeing.	

Funky Flavour upcycled cap by Pleasant. A cotton bed sheet from a local charity in Denmark was simply washed, then converted directly into a cap.

pleasant,

Recycled wool textiles

Wool recycling has a long history, and several suppliers offer high-quality materials with a significantly reduced environmental footprint compared with virgin wool.

Suppliers & materials	- Manteco® MWool® recycled wool - Kvadrat Re-wool recycled wool	
Raw material origin	Typically a combination of PIR and PCR wool textile waste, often mixed with virgin material. Ask suppliers to provide ISCC PLUS, SCS Recycled Content or other certification that verifies the ratio and type of recycled content.	
	Manteco® MWool® recycled wool[14]	**Virgin wool**[15]
GWP	0.6 kg CO_2e / kg fibre	10.7 kg CO_2e / kg fibre
Energy use	8.4 MJ / kg fibre	8.7 MJ / kg fibre
Water use	90 l / kg fibre	38.2 l / kg fibre
GWP	3.2 kg CO_2e / kg woven textile (500 dtex)	16.1 kg CO_2e / kg woven textile (500 dtex)
Toxicity	Pure wool textiles are considered non-toxic, but certain ingredients used in textile finishing processes have been linked to toxicity. Request a material safety data sheet for dyes, colourants and other finishes that are used with specific recycled wool materials and confirm REACH compliance and other relevant certification with suppliers.	
Circularity	Recycled wool textiles accounted for around 6% of the total 1 million tonnes of wool textiles produced globally in 2020.[16] Avoid combinations with other types of textile fibre to increase the likelihood of recycling on disposal. Pure wool textiles are compostable at home or industrially. For an overview of recycling rates and circular design guidelines for textiles, see page 75.	
Mechanical properties	Very durable and hardwearing, with excellent abrasion resistance, although durability and toughness depend on the thickness of yarn and the process used – woven textiles tend to be stronger than knits, for example.	
Environmental resistance	Good breathability; good UV and thermal resistance.	
Forming	Weaving and knitting, as well as felting and other non-woven processes. The material can be further formed using common textile processes such as cutting and sewing.	
Finishing	The colour of recovered wool textile waste will impact the colourability of the recycled material, although careful sorting by colour, overdyeing and mixing with virgin material can allow suppliers to offer a wider range of colour options. Compatible with many textile finishing processes, including printing and dyeing.	

Arbour Club Armchair, designed by Andreas Engesvik and Daniel Rybakken for HAY. The upholstery textile is made with Re-wool, a 45% recycled wool textile from Kvadrat.

Recycled leather

Recycled leather materials – often called bonded leathers – are typically made with post-industrial leather waste that is ground up and mixed with a resin binder or reconstituted using a **fibre entanglement process**. Depending on the specific process, recycled leathers can be very similar to virgin material in terms of performance and aesthetics.

Suppliers & materials	- ELeather recycled leather - Prodotti Alfa CORIUM® recycled leather - Spinnova/KT Trading Respin recycled leather	
Raw material origin	Mostly PIR leather waste, and to a lesser extent, PCR leather. Ask suppliers to confirm the ratio and type of recycled content, as well as detailed information about any resin binders or composite structure used for reconstituted leather.	
	Recycled leather[17]	**Virgin leather**[18]
GWP	16.6 kg CO_2e / kg	41.5 kg CO_2e / kg
Energy use	No data	335 MJ / kg
Water use	No data	32,800 l / kg
Toxicity	Leather itself is non-toxic and the use of chemicals and other potentially harmful ingredients in leather tanning is strictly regulated in key markets like the EU. Confirm REACH and SVHC (Substances of Very High Concern) compliance and other relevant certification with suppliers.	
Circularity	Although leather recycling statistics were unavailable at the time of writing, it's estimated that the global leather industry generates around 800,000 tonnes of PIR leather waste annually.[19] Many leather suppliers have a zero-waste policy, with uses for leather waste as a raw material in fertilizer, as well as in the pharmaceutical and cosmetics industries, to name just a few. Specialist recyclers like the ones featured here are able to recycle PIR and PCR leather waste into recycled leather materials. Pure leather is biodegradable, but tanned leather usually isn't. Some suppliers offer tanning processes that result in biodegradable leather, such as Zeology, by the Dutch supplier Nera.	
Mechanical properties	Properly cared for, leather can be very durable, with good strength and abrasion resistance.	
Environmental resistance	Good moisture and thermal resistance, with modest UV resistance.	
Forming	Compatible with common textile forming and assembly processes such as cutting and sewing, as well as compression moulding for three-dimensional shapes.	
Finishing	The colour of recovered leather waste will impact the colourability of the recycled material, although careful sorting by colour, overdyeing and mixing with virgin material can allow suppliers to offer a wider range of colour options. Leather can be finished using dyeing, printing, debossing/embossing and laser marking, to name a few compatible processes.	

Wallet and powerbank by INÉ.
The outer cover is made with PIR leather waste.

INÉ

Renewable textiles

According to Textile Exchange's 2021 *Preferred Fiber and Materials Market Report*, renewable textiles accounted for an estimated 37 per cent of the total global production of textiles in 2020. This family of materials can be subdivided into three groups – plant-based natural textiles, animal-based textiles and renewable synthetic textiles.

All of these materials have their strengths and weaknesses from an environmental perspective. Within the plant-based group, some – like hemp and jute – can grow on non-agricultural land using minimal irrigation, whereas cotton requires good soil and plenty of water, and many cotton growers use large amounts of petrochemical-based fertilizers and pesticides. Several initiatives aim to reduce the environmental impact of cotton, such as organically grown cotton that doesn't use petrochemical-based fertilizers or pesticides at all, as well as Better Cotton, which aims to eliminate the worst pollutants and reduce the use of fertilizers and pesticides.

Other factors affect animal-based textiles, including emissions associated with cattle farming, animal welfare and, in extreme cases, deforestation and soil erosion when land is cleared for grazing animals. Textile Exchange's Responsible Wool Standard requires farmers to manage grazing pastures to avoid soil degradation, as well as using responsible wool-shearing practices. The OEKO-TEX® Leather Standard and Leather Working Group (LWG) certification require suppliers to fulfil certain criteria around animal welfare, traceability of hides, and energy, water and chemical usage.

Cellulose-based synthetic textiles belong in the third group. While the cellulose raw material is derived from timber and, to a lesser extent, cotton, converting cellulose fibre to synthetic filament for textile production is energy-intensive and, in the case of viscose, typically uses harsh chemicals. Canopy, an NGO set up to protect the world's forests, estimates that a third of global cellulose-based synthetic textile production uses timber from illegal logging, highlighting the importance of confirming that suppliers source only FSC- or PEFC- or similarly certified timber. Other suppliers are developing new processes for cellulose-based synthetics textile production, including Circular Systems' Agraloop™ process, which uses agricultural waste.

Lastly, this third group also includes several renewable alternatives to conventional petrochemical-based synthetic textiles, such as renewable polyester (PET), polyamide (PA) and polylactide (PLA) textiles. These face some of the same environmental challenges around the origin of raw materials as renewable plastics. Renewable PET and PLA textiles are typically made with sugar from food crops like corn and beets that are grown on agricultural land. Volumes are still small, so production doesn't have a major impact on food production for now, but this could change if volumes increase. Other renewable synthetic textiles, like renewable PA, are made with non-food crops that don't need to be grown on agricultural land. And beyond these examples, processes are also being developed for making textiles with alternative renewable raw materials such as algae (see pages 94–97).

Renewable textile raw materials

A wide range of renewable raw materials are suitable for textile production. The environmental impact of these varies considerably, depending on growing conditions, the potential impact on food production and the manufacturing steps needed to convert the raw materials into yarns and textiles.

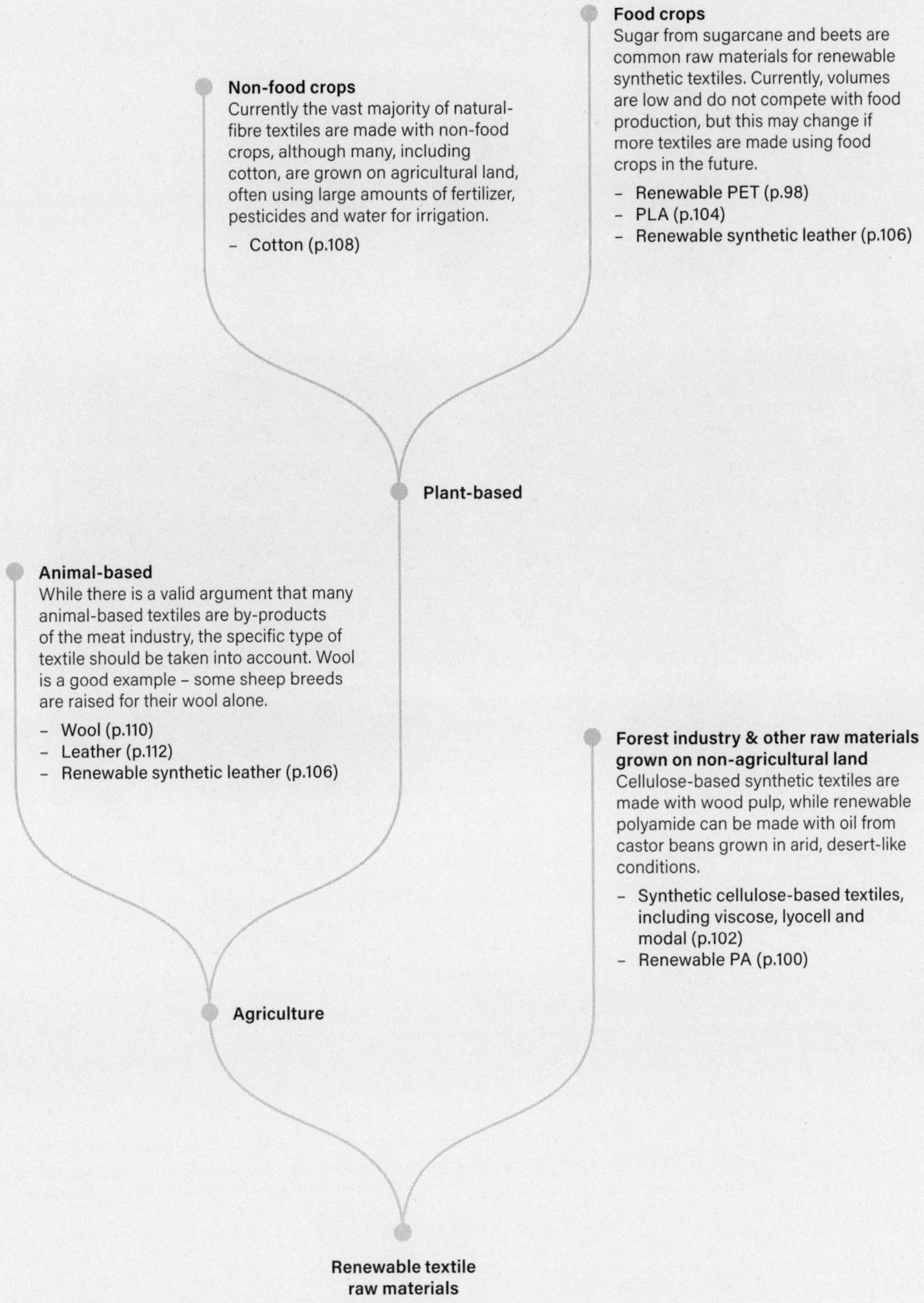

Food crops

Sugar from sugarcane and beets are common raw materials for renewable synthetic textiles. Currently, volumes are low and do not compete with food production, but this may change if more textiles are made using food crops in the future.

- Renewable PET (p.98)
- PLA (p.104)
- Renewable synthetic leather (p.106)

Non-food crops

Currently the vast majority of natural-fibre textiles are made with non-food crops, although many, including cotton, are grown on agricultural land, often using large amounts of fertilizer, pesticides and water for irrigation.

- Cotton (p.108)

Plant-based

Animal-based

While there is a valid argument that many animal-based textiles are by-products of the meat industry, the specific type of textile should be taken into account. Wool is a good example – some sheep breeds are raised for their wool alone.

- Wool (p.110)
- Leather (p.112)
- Renewable synthetic leather (p.106)

Forest industry & other raw materials grown on non-agricultural land

Cellulose-based synthetic textiles are made with wood pulp, while renewable polyamide can be made with oil from castor beans grown in arid, desert-like conditions.

- Synthetic cellulose-based textiles, including viscose, lyocell and modal (p.102)
- Renewable PA (p.100)

Agriculture

Renewable textile raw materials

'We need to ask ourselves what it's going to feel like to live in a post-carbon society.'

An interview with Charlotte McCurdy, an industrial designer and researcher whose work bridges the gap between design, material technology and sustainable innovation. Besides being an assistant professor at Arizona State University, Charlotte runs a design consultancy and materials lab in New York. Find out more at charlottemccurdy.com

The scope of your work and your thinking about sustainability is both original and refreshing. Could you tell me a bit about your background and how you became interested in sustainability in the first place?

I remember a presentation at school when I was eight years old, about the risk of running out of potable water, which really opened my eyes to the effects that unsustainable systems are having on the environment. But that presentation also made me optimistic, in that it made me see that change is possible if we can focus our imagination on how to make things better and organize a large-scale response based on that.

Later, at the beginning of my career, I worked as a sustainability consultant at an NGO, doing a lot of environmental analysis and programme design for large food and beverage companies. Rather than working on a shared vision for the benefit of everyone, I felt like I was helping to formalize ideas about sustainability and creating structures for companies to get away with changing as little as possible. My work there also made me think about the relationship between the very large companies that I was consulting for and the small farmers that they source their raw materials from. Thinking about these relationships has given my approach to sustainability a strong humanist bent, because how can we have empathy for animals and living systems if we don't yet demonstrate adequate empathy for other humans?

Algae grown in the Charlotte McCurdy lab.

This got me into the more conceptual side of how sustainability could be defined, which in turn got me into design as a space where these questions can be explored at the intersection of technological solutions, political will and the collective imagination.

I've always struggled with the side of the environmental movement that seems to be saying that the best thing for the environment would be if us humans could just disappear from the face of the earth. By contrast, you seem much more focused on finding pragmatic and holistic solutions.

I try to tell an optimistic story about carbon-negative materials and decarbonization because that's how I can get people to listen to me for 30 seconds longer. I know this isn't the whole story – we absolutely do over-consume and have a throwaway culture, and that needs to change – but going to the other extreme and saying that people should just wear their clothes forever, or use their furniture forever, is just as unreasonable because entropy is real. I don't think that fighting climate change means going back to what we had before the Industrial Revolution. It's going to look completely different from that. We really need to ask ourselves what it's going to feel like to live in a post-carbon society.

Your work with algae-based materials and how they fit into the bigger picture of the carbon cycle is really interesting. What set you off in this direction?

I knew I had to explore algae because they are so efficient at converting solar energy into useful molecules. Photosynthetic efficiency refers to the proportion of energy from sunlight that plants convert and store as chemical energy during photosynthesis. The most efficient land-based plants have a photosynthetic efficiency of about 2 per cent, while algae have 6 to 8 per cent and therefore greater potential for sequestering carbon dioxide.

The carbon-negative raincoat that I designed is embodied CO_2, and the subsequent question is, what's the best thing that could happen to it at the end of its life? My provocative response, which sometimes gets people wound up, is that if we're going to prioritize fighting against climate change, it should probably go into modern landfill, because this way its CO_2 would be stored away. I think of it as consumer-driven carbon sequestration, and we should see it as a complement to other approaches, like trying to regulate emissions through legislation and taxes.

Carbon-negative raincoat, made with an algae-based plastic material developed in the Charlotte McCurdy lab.

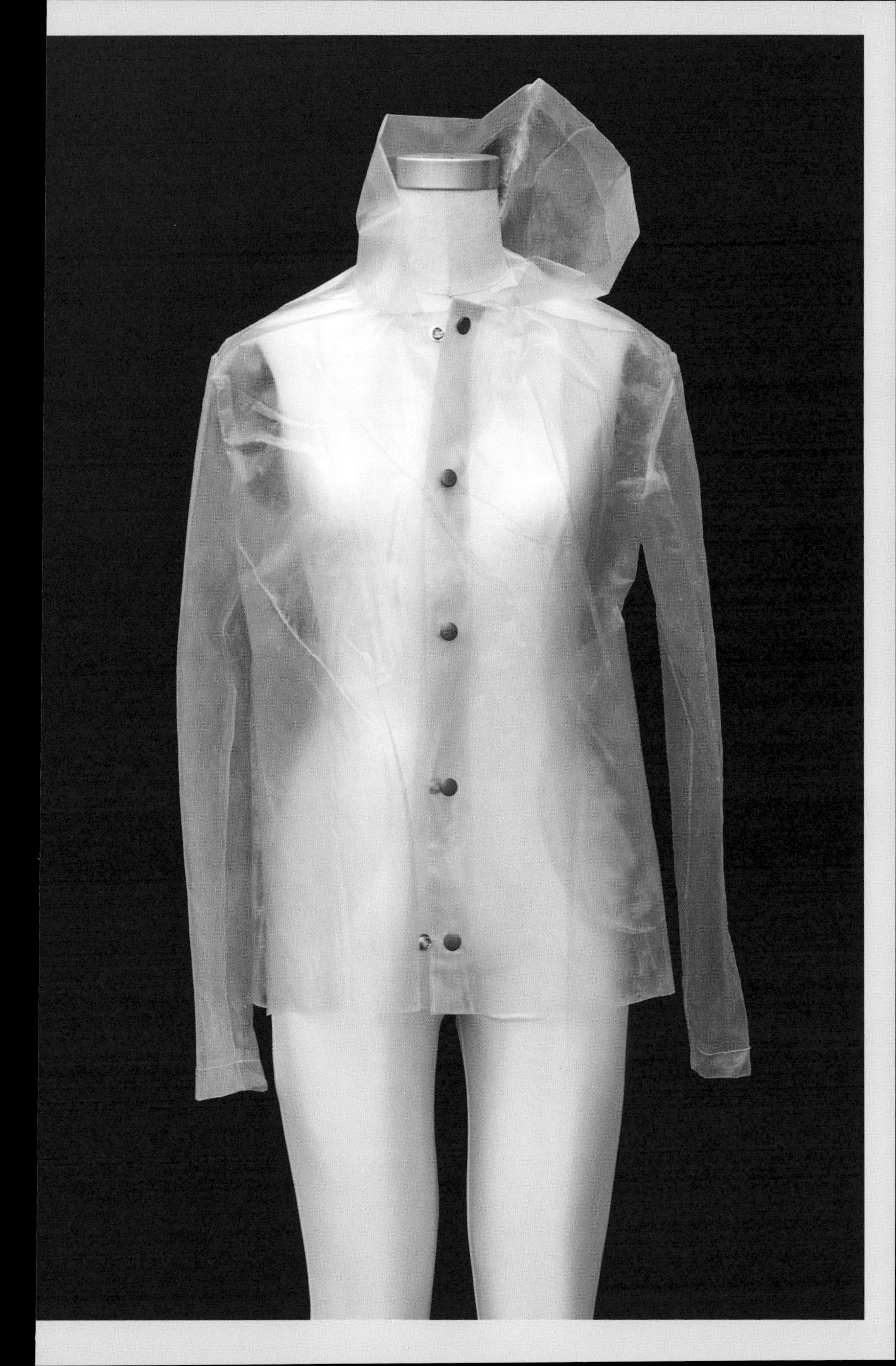

Renewable polyester (PET) textiles

Currently, it's possible to replace some of the petrochemical-based ingredients in PET filament production with renewable alternatives, with the aim being to enable fully renewable PET textiles in the future.

Suppliers & materials

- Virent BioForm PX® renewable PET filament
- Toray Ecodear® renewable PET filament
- Radici Biofeel® PET filament

Raw material origin

While fully renewable PET filament has been successfully produced at small scale, most commercially available renewable PET filament consists of about 30% renewable content. Raw materials include sugar derived from food crops such as sugarcane, beets and corn, but also agricultural waste, grass and wood. Ask suppliers to provide Bonsucro, ISCC PLUS or other certification that verifies the origin and ratio of renewable raw materials.

	Virent BioForm PX® renewable PET filament[1]	Virgin petrochemical-based PET textile[2]
GWP	2.2 kg CO_2e / kg fibre	2.7 kg CO_2e / kg fibre
Energy use	No data	78.4 MJ / kg fibre
Water use	No data	No data
GWP	5.2 kg CO_2e / kg woven textile (500 dtex)	5.7 kg CO_2e / kg woven textile (500 dtex)

Toxicity

Washing PET textiles is believed to be a major cause of the spread of plastic microfibres, but other than this, pure PET textiles are considered non-toxic. Certain ingredients used in textile finishing processes have been linked to toxicity. Request a material safety data sheet for dyes, colourants and other finishes, and confirm REACH compliance and other relevant certification with suppliers.

Circularity

Renewable PET textiles can be recycled with other PET textile waste without separation. While very few PET textiles are recycled (see page 80), the likelihood of recycling can be improved by using pure PET textiles, rather than blends with other textile fibres. Renewable PET textiles are not biodegradable. For an overview of recycling rates and circular design guidelines for textiles, see page 75.

Mechanical properties

Strong and durable, yet lightweight and flexible. Weaving tends to result in stronger, more durable textiles, while knitted textiles offer more flexibility. PET felt gives good cushioning.

Environmental resistance

Water resistance is good, although PET textiles tend to absorb grease, making stain removal difficult. Unlike natural textiles, PET textiles are not breathable.

Forming

Weaving and knitting, as well as felting, artificial down and other non-woven processes. The material can be further formed using common textile processes such as cutting, sewing and ultrasonic welding.

Finishing

Dyeing is straightforward and PET textiles are compatible with many textile finishing processes, including printing, embossing and debossing, transfers and decorative lamination, although the latter should be PET-based so as not to impact circularity.

Slingsby Ultra Jacket by Bergans. The woven outer textile is made with Ecodear®, partially renewable PET filament.

Bergans

Renewable polyamide (PA) textiles

Renewable PA (or nylon) textiles are typically derived from castor beans, a non-food crop that can grow on arid, non-agricultural land. This tough and durable material is available from several suppliers in many different grades.

Suppliers & materials	- Fulgar EVO® renewable PA 6.6 filament - Swicofil renewable PA 11 filament - Radici Biofeel® PA 6.10 filament	
Raw material origin	Typically between 70 and 80% renewable content. The dominant raw material is oil derived from castor beans. Ask suppliers to provide ISCC PLUS or other certification that verifies the origin and ratio of renewable content.	
	Fulgar EVO® renewable PA 6.6 filament[3]	**Virgin petrochemical-based PA6 filament**[4]
GWP	7.4 kg CO_2e / kg fibre	9.6 kg CO_2e / kg fibre
Energy use	No data	148 MJ / kg fibre
Water use	8,640 l / kg fibre	No data
GWP	10.3 kg CO_2e / kg woven textile (500 dtex)	12.6 kg CO_2e / kg woven textile (500 dtex)
Toxicity	As with PET textiles, washing PA textiles is believed to be a source of microfibre proliferation, although to a lesser degree. Pure PA textiles are considered non-toxic, but certain ingredients used in textile finishing processes have been linked to toxicity. Request a material safety data sheet for dyes, colourants and other finishes, and confirm REACH compliance and other relevant certification with suppliers.	
Circularity	Renewable PA textiles can be recycled with petrochemical-based PA textiles without separation. However, it's preferable to separate different types of PA for optimal results, as mixing different types (i.e. PA 6 with PA 11) can give unpredictable results. Avoid combinations with other types of textile fibre to increase the likelihood of recycling on disposal. Renewable PA textiles are not biodegradable. For an overview of recycling rates and circular design guidelines for textiles, see page 75.	
Mechanical properties	Tough and durable, although the specific type of fabric will have an impact on performance - woven PA textiles can be extremely tough and hardwearing, for example, while knitted PA textiles are generally more elastic and less durable.	
Environmental resistance	While PA11 has good moisture resistance, PA6 and 6.6 textiles readily absorb moisture, so odours are potentially retained. Generally, PA textiles are also sensitive to heat, with a melting point of around 100°C (200°F).	
Forming	Weaving and knitting, as well as felting and other non-woven processes. The material can be further formed using common textile processes such as cutting, sewing and ultrasonic welding.	
Finishing	Dyeing is straightforward and PA textiles are compatible with many textile finishing processes, including printing, embossing and debossing, transfers and decorative lamination, although the latter should be PA based so as not to impact circularity.	

Tekoa Biobased Pants by Vaude. The textile used in the trousers is made with renewable polyamide yarn derived from castor beans, a non-food crop that doesn't require agricultural land.

Cellulose-based synthetic textiles

Cellulose-based synthetic textiles are made with so-called dissolving pulp, which is mostly derived from timber and, to a smaller extent, short cotton fibre not used in cotton textile production. Common cellulose-based synthetic textiles include viscose, lyocell and modal.

Suppliers & materials	- Sateri EcoCosy® viscose fibre - Metsä Kuura cellulose-based synthetic fibre - Birla Cellulose™ Livaeco modal filament
Raw material origin	Most cellulose-based synthetic textile production is timber-based, specifically eucalyptus, spruce and pine, but bamboo and cotton can also be used. Ask suppliers to provide FSC, PEFC or other certification that verifies the origin of raw materials.
	Virgin viscose fibre[5]
GWP	1.5 kg CO_2e / kg fibre
Energy use	20.6 MJ / kg fibre
Water use	No data
GWP	3.2 kg CO_2e / kg woven textile (500 dtex)
Toxicity	There are four basic types of cellulose-based synthetic textile fibres – viscose (or rayon), lyocell, modal and cupro. Various chemicals are used in the fibre manufacturing process, including some that are hazardous. The use of these chemicals can be very harmful to factory workers and the local environment unless properly managed. Request a material safety data sheet for specific fibres, dyes, colourants and other finishes, and confirm REACH compliance and relevant certification with suppliers.
Circularity	Recycled textiles accounted for around 0.4% of the total 6.5 million tonnes of cellulose-based synthetic textiles produced globally in 2020.[6] This is expected to grow as chemical recycling processes become more established. Avoid combinations with other types of textile fibre to improve the likelihood of recycling on disposal. Some cellulose-based synthetic textiles are biodegradable; ask suppliers for confirmation and relevant certification. For an overview of recycling rates and circular design guidelines for textiles, see page 75.
Mechanical properties	Typically soft and rather delicate, although durability and toughness depend on the yarn count and textile construction. These fibres can also be used for very robust technical textiles.
Environmental resistance	Breathable and absorbent. Cellulose-based synthetic textiles have poor resistance to hot water and UV radiation, but some suppliers offer specially treated textiles with improved UV resistance.
Forming	Weaving and knitting, as well as felting and other non-woven processes. The material can be further formed using common textile processes such as cutting and sewing.
Finishing	Dyeing cellulose-based synthetic textiles can be more complicated than with other textiles, as some materials are easily damaged during the process. Other finishing processes include printing.

Tree-Kånken by Fjällräven. The bag is made with lyocell, one of the major types of cellulose-based synthetic textile.

FJÄLLRÄVEN
KÅNKEN

Polylactide (PLA) textiles

Like rigid PLA plastic, PLA textiles are derived from sugar from plants. PLA textiles may not have the highest performance among synthetic textiles, but they can be a renewable alternative with similar properties to recycled and renewable PET textiles (see pages 80–81 and 98–99).

Suppliers & materials	- NatureWorks Ingeo PLA filament - ADVANSA ADVA® Shortcut PLA filament
Raw material origin	Typically made entirely with sugar derived from plants like sugarcane and beets. Ask suppliers to provide Bonsucro, OK Biobased or other certification that verifies the origin of raw materials.
	NatureWorks Ingeo PLA filament[7]
GWP	3.4 kg CO_2e / kg fibre
Energy use	67.7 MJ / kg fibre
Water use	35 l per / kg fibre
GWP	3.6 kg CO_2e / kg woven textile (500 dtex)
Toxicity	Pure PLA textiles are considered non-toxic, but some textile finishing processes use toxic ingredients. Request a material safety data sheet for dyes, colourants and other finishes, and confirm REACH compliance and other relevant certification with suppliers.
Circularity	Currently not widely recycled, most likely because only a few hundred thousand tonnes of PLA textiles are produced annually – although volumes are growing.[8] Avoid combinations with other types of textile fibre to increase the likelihood of recycling on disposal. Pure PLA textiles are compostable at home or industrially. For an overview of recycling rates and circular design guidelines for textiles, see page 75.
Mechanical properties	Similar properties to PET textiles – strong and durable, but with marginally worse abrasion resistance. Durability and toughness also depend on the thickness of yarn and the process used – woven textiles tend to be stronger than knits, for example.
Environmental resistance	Unlike PET textiles, PLA textiles are breathable. They also offer good UV and stain resistance, but lower thermal resistance than PET.
Forming	Weaving and knitting, as well as felting and other non-woven processes. The material can be further formed using common textile processes such as cutting, sewing and ultrasonic welding.
Finishing	Dyeing PLA textiles is relatively difficult, as only specific dyes are compatible and conditions need to be carefully controlled. Other finishing processes include printing, embossing and debossing, transfers and decorative lamination, although the latter should be PLA-based so as not to impact circularity.

Non-woven PLA textile in a tea bag. So far, the majority of commercially launched products that use PLA textiles have been in packaging and technical textiles for industrial applications.

Renewable synthetic leather

A new generation of synthetic leather materials is derived from plant- and animal-based protein grown in labs. Currently, commercially available materials are typically composites that consist of lab-grown materials, renewable polymers and a textile backing, while future technologies will be fully lab-grown.

Suppliers & materials	- BioFabbrica Bio-Tex™ renewable synthetic leather - VitroLabs renewable synthetic leather	
Raw material origin	BioFabbrica Bio-Tex™ is made with a combination of lab-grown protein-based artificial leather, renewable polyurethane (PU) elastomer and a textile backing. Alternative technologies such as VitroLabs are fully lab-grown. Ask suppliers for detailed information about the material structure and ingredients, as well as ISCC PLUS or any other relevant certification verifying the origin and ratio of renewable content.	
	BioFabbrica Bio-Tex™ renewable synthetic leather[9]	**Virgin leather**[10]
GWP	8.2 kg CO_2e / kg	41.5 kg CO_2e / kg
Energy use	158 MJ / kg	335 MJ / kg
Water use	98 l / kg	32,800 l / kg
Toxicity	Unlike many conventional artificial leathers, the materials listed here do not contain PVC. BioFabbrica Bio-Tex™ is partially made with polyurethane (PU), however. There is currently no clear link between PU materials and toxic VOCs during normal use, but the material will give off toxic fumes when burnt. Request a material safety data sheet for specific materials and confirm REACH compliance and other relevant certification with suppliers.	
Circularity	Synthetic leathers, including the materials listed here, are not widely recycled, as they typically consist of a multilayered composite structure that's difficult to separate and recycle. For an overview of recycling rates and circular design guidelines for textiles, see page 75.	
Mechanical properties	Overall high strength, with good tear and abrasion resistance.	
Environmental resistance	Good weatherability; good moisture and chemical resistance.	
Forming	Compatible with common textile forming and assembly processes such as cutting, sewing and ultrasonic welding, as well as compression moulding for three-dimensional shapes.	
Finishing	Typically available in a range of standard colours, with colour matching for larger orders. Compatible with many textile finishing processes, including printing, embossing/debossing and laser marking.	

New Day Market Tote by Everlane. The bag is made with BioFabbrica Bio-Tex™, a partly lab-grown renewable synthetic leather material.

Cotton textiles

Virgin cotton has a sizeable environmental footprint, mainly associated with irrigation, petrochemical-based fertilizers and pesticides. Certified organic cotton and cotton that is grown using standards that limit the use of water and chemicals can offer significant reductions.

Suppliers & materials	Free tools for sourcing organic and other responsibly produced cotton include OEKO-TEX®'s Buying Guide, the GOTS Certified Suppliers Database and Textile Exchange's supplier database.	
Raw material origin	Virgin cotton grown organically or with limited use of fertilizers and pesticides, and other criteria aimed at reducing the environmental impact. Ask suppliers to provide GOTS, OEKO-TEX®, Fairtrade Organic, Better Cotton or other certification to confirm specific growing conditions.	
	Organic virgin cotton[11]	**Virgin cotton**[12]
GWP	1 kg CO_2e / kg fibre	1.8 kg CO_2e / kg fibre
Energy use	5.8 MJ / kg fibre	15 MJ / kg fibre
Water use	180 l / kg fibre	2,100 l / kg fibre
GWP	3.5 kg CO_2e / kg woven textile (500 dtex)	4.3 kg CO_2e / kg woven textile (500 dtex)
Toxicity	Organic cotton doesn't allow genetically modified (GMO) seeds and is grown without pesticides and petrochemical-based fertilizers, while other initiatives such as Better Cotton allow GMO seeds while limiting the use of fertilizers and pesticides. While pure cotton textiles are considered non-toxic, some textile finishing processes use toxic ingredients. Request a material safety data sheet for specific dyes, colourants and other finishes, and confirm REACH compliance and other relevant certification with suppliers.	
Circularity	Recycled cotton textiles only accounted for around 0.96% of the total 26.2 million tonnes of cotton textiles produced globally in 2020.[13] Avoid combinations with other types of textile fibre to increase the likelihood of recycling on disposal. Pure cotton textiles are compostable at home or industrially. For an overview of recycling rates and circular design guidelines for textiles, see page 75.	
Mechanical properties	Soft and rather delicate, although durability and toughness depend on the thickness of yarn and the process used – woven textiles tend to be stronger than knits, for example.	
Environmental resistance	Poor UV resistance, so colours will typically fade over time. Good breathability and thermal resistance, although cotton textiles may shrink when washed at high temperatures.	
Forming	Weaving and knitting, as well as felting and other non-woven processes. The material can be further formed using common textile processes such as cutting and sewing.	
Finishing	Compatible with many textile finishing processes, including dyeing and printing.	

Flamingo pink speaker cover, designed by Unisk for the SYMFONISK speaker by IKEA. The cover is made with a woven textile that consists of a blend of virgin organic and recycled cotton.

Wool textiles

Virgin wool textiles have a large environmental footprint, mainly associated with livestock farming, land use and animal welfare. Several initiatives and organizations aim to reduce the impact of wool production; see below for a top-level overview.

Suppliers & materials	Free tools for sourcing responsibly produced wool include OEKO-TEX®'s Buying Guide, the GOTS Certified Suppliers Database and Textile Exchange's supplier database.
Raw material origin	Virgin wool fibre, mainly from merino sheep. Textile Exchange's Responsible Wool Standard (RWS) requires farmers to manage grazing pastures to avoid soil degradation and forbids mulesing and other practices that are cruel to animals. Ask suppliers to provide RWS or other relevant certification.
	Virgin wool[14]
GWP	75.8 kg CO_2e / kg fibre
Energy use	125 MJ / kg fibre
Water use	13,900 l / kg fibre
GWP	78.3 kg CO_2e / kg woven textile (500 dtex)
Toxicity	Pure wool textiles are considered non-toxic, but some textile finishing processes use toxic ingredients. Request a material safety data sheet for specific dyes, colourants and other finishes, and confirm REACH compliance and relevant certification with suppliers.
Circularity	Recycled wool textiles accounted for around 6% of the total 1 million tonnes of wool textiles produced globally in 2020.[15] Avoid combinations with other types of textile fibre to increase the likelihood of recycling on disposal. Pure wool textiles are compostable at home or industrially. For an overview of recycling rates and circular design guidelines for textiles, see page 75.
Mechanical properties	Very durable and hardwearing, with excellent abrasion resistance, although durability and toughness depend on the thickness of yarn and the process used – woven textiles tend to be stronger than knits, for example.
Environmental resistance	Good breathability; good UV and thermal resistance.
Forming	Weaving and knitting, as well as felting and other non-woven processes. The material can be further formed using common textile processes such as cutting and sewing.
Finishing	Compatible with many textile finishing processes, including dyeing and printing.

OSLO speaker by VIFA,
with a wool textile cover.

Leather

While the environmental impact of livestock farming can be considerable, leather is essentially a by-product of the meat and dairy industry, as well as an exceptionally durable and useful material in its own right.

Suppliers & materials	Free tools for sourcing responsibly produced leather include the Leather Working Group (LWG)'s certified supplier list and OEKO-TEX®'s Buying Guide. Textile Exchange's supplier database covers sheepskin and leather.
Raw material origin	Virgin leather from cattle, sheep and other animals. Several organizations are working for more sustainable practices in the leather industry based on animal welfare, traceability of hides, and energy, water and chemical usage. Ask suppliers to provide LWG, OEKO-TEX® Leather Standard or other relevant certification.
	Virgin leather[16]
GWP	41.5 kg CO_2e / kg
Energy use	335 MJ / kg
Water use	32,800 l / kg
Toxicity	Leather itself is non-toxic, and the use of chemicals and other potentially harmful ingredients in leather tanning is strictly regulated in key markets like the EU. Many leather suppliers clean and recycle the water used in production. Confirm REACH and SVHC (Substances of Very High Concern) compliance and relevant certification with suppliers.
Circularity	Many leather suppliers have a zero-waste policy, with uses for leather waste as a raw material in fertilizer, as well as in the pharmaceutical and cosmetics industries, to name just a few examples. Specialist recyclers are also able to recycle PIR and PCR leather waste into recycled leather materials (see pages 90–91). Pure leather is biodegradable, but tanned leather usually isn't. Some suppliers offer tanning processes that result in biodegradable leather, such as Zeology, by the Dutch supplier Nera.
Mechanical properties	Properly cared for, leather can be very durable, with good strength and abrasion resistance.
Environmental resistance	Good moisture and thermal resistance, with modest UV resistance.
Forming	Compatible with common textile forming and assembly processes such as cutting and sewing, as well as compression moulding for three-dimensional shapes.
Finishing	Compatible with many textile finishing processes, including dyeing, printing, embossing/debossing and laser marking.

Space Ovals, created by Sørensen Leather in collaboration with Space Copenhagen, and upholstered in Sørensen Leather's NUANCE leather.

3 Metals

Metal production is very energy-intensive and the industry is a major source of pollution – according to a 2021 Global Energy Monitor report, just 553 steel plants were responsible for 9 per cent of global CO_2 emissions that year. And while aluminium is produced in much smaller volumes than steel, it's an even more energy-intensive process – emissions can be as high as 20 kilos of CO_2 for a single kilo of material. Comparing the environmental footprint of metals can be more complicated than this, however. Although steel and aluminium are often viable replacements for each other, a steel part will likely look very different to an aluminium part, even if designed for the same application – partly because forming processes can vary a lot between the two, but the density of steel is also about three times that of aluminium; a steel part will weigh about three times more than an aluminium part of the same volume. The common way of measuring GWP using kilos of CO_2e per kilo of material can therefore be misleading when comparing metals, as shown in the diagram opposite.

Mining is another aspect of the metals industry that can have profoundly negative environmental effects, such as accidents with contaminated waste water leaching into local soil and waterways. The Initiative for Responsible Mining Assurance (IRMA) awards certification to mining companies that fulfil certain criteria for responsible mining, while the Aluminium Stewardship Initiative (ASI), ResponsibleSteel™ and others audit and award certification for each stage of the manufacturing process, from mining to finished product.

On the other hand, metals are extremely durable and widely recycled, due to their high value and the relative ease with which they can be separated from other waste. The International Aluminium Institute estimates that about 75 per cent of the 1.5 billion tonnes of aluminium that has ever been produced is still in use. On average, about 35 per cent of the steel and 32 per cent of the aluminium in global production is recycled material.

This chapter looks at a range of steel and aluminium materials that use a high ratio of recycled content, as well as processes for manufacturing low-carbon metals. I would argue that steel and aluminium are the dominant metals in product design. Others, such as copper, zinc, titanium, precious metals and rare earth metals, are also important raw materials, but beyond the scope of this book.

Global annual demand and emissions from metal production

The metal industries are a major source of CO_2 emissions, with an estimated 500 steel mills across the globe responsible for about 9% of total emissions. Aluminium production is very energy-intensive, with emissions at about a third of those of steel, although annual aluminium production volumes are about 30 times lower.[1]

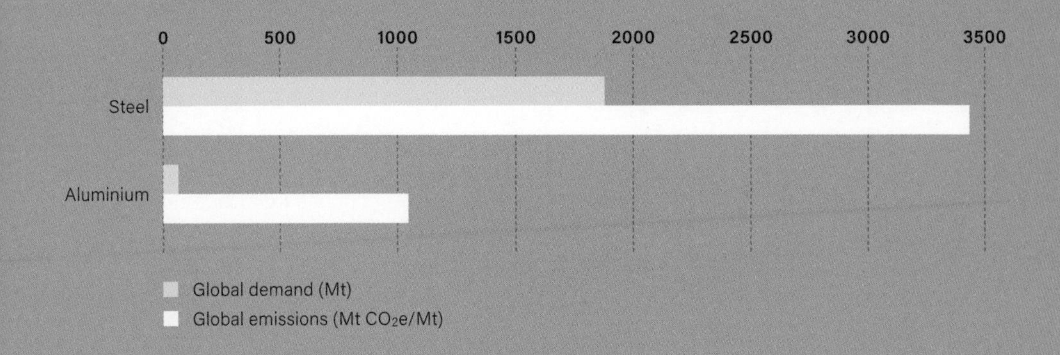

Benchmarking the environmental impact of metals

Looking at emissions based on kilos of CO_2 per kilo of material is a common mistake when comparing the environmental impact of metals. The density of steel is about three times that of aluminium, so using a reference volume like a cubic metre or a specific part design will give a more accurate result.[2]

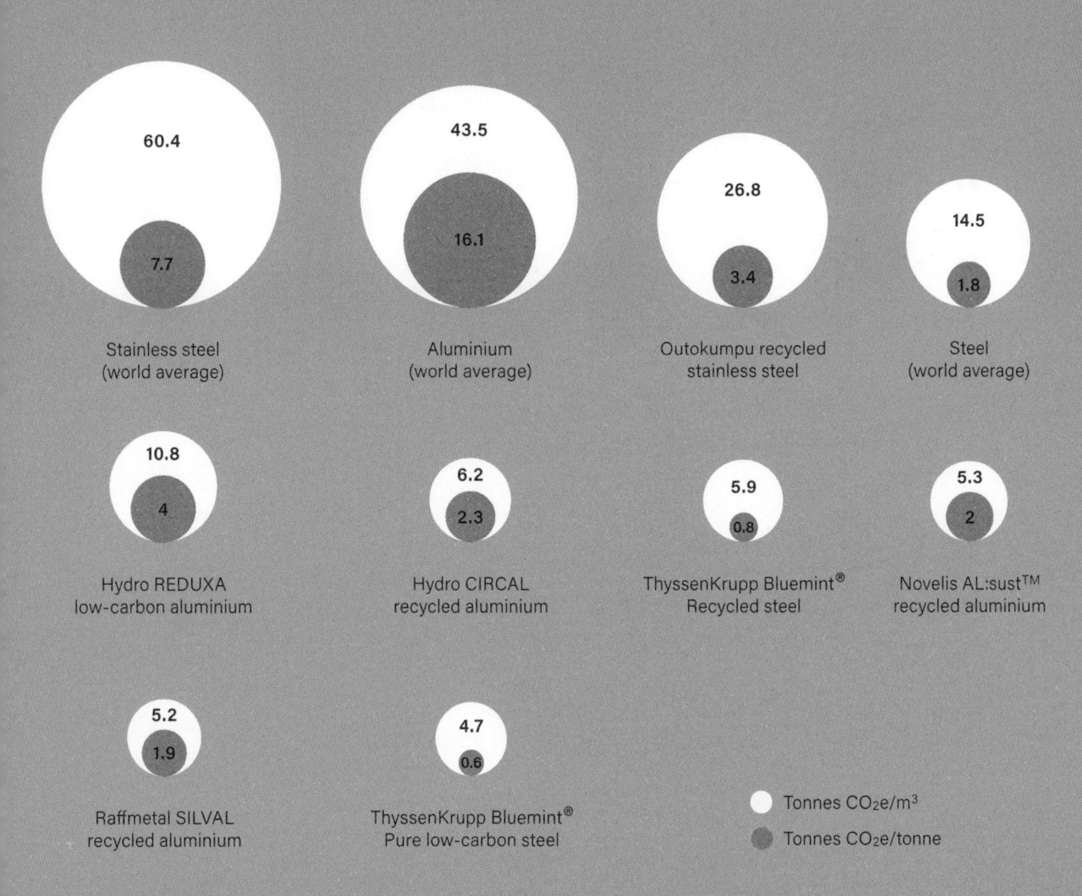

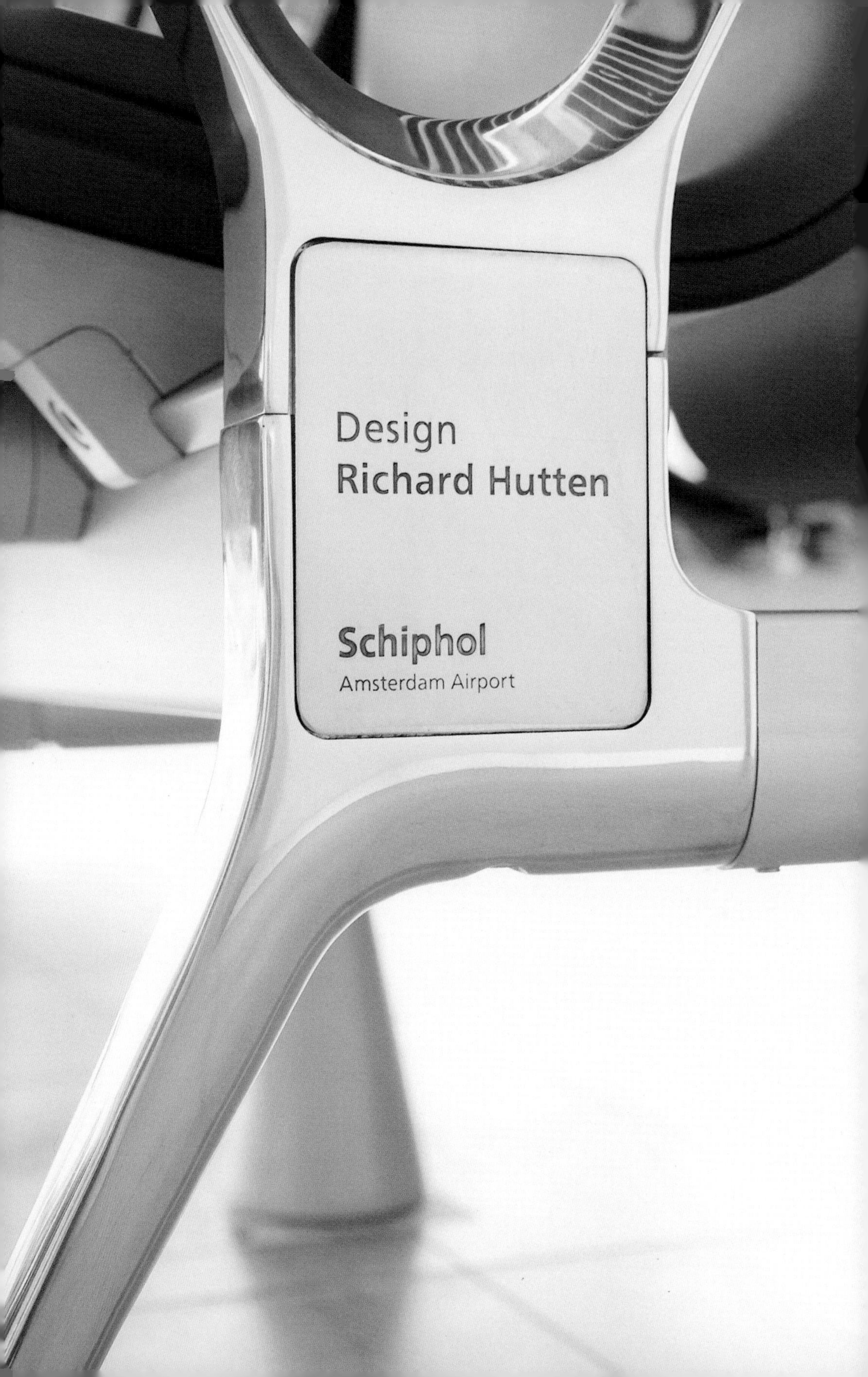
Design
Richard Hutten
Schiphol
Amsterdam Airport

'I'm a designer, so I'm interested in materials.'

An interview with Richard Hutten, who in 2020 completed a redesign of 27,000 aluminium seats at Schiphol Airport by recycling the airport's existing aluminium furniture and manufacturing the new pieces locally. Find out more about his work at richardhutten.com

I wanted to start by asking if there was an initial 'spark' that made you interested in sustainability as a designer?

At the beginning of my career, I used a lot of wood, which of course today is considered a very sustainable material, being renewable and removing carbon from the air while it's growing. But at the time I was shocked to see workers in factories wearing big masks while spraying wooden furniture with coatings that made the material look like plastic. The natural quality of the material was lost and clearly this process was not healthy for the workers. So I researched wood finishes and started using linseed oil, which earned me a Dutch Furniture Award in 1991 and sparked a trend among interior companies for using traditional oils, waxes and other natural wood finishes. As a designer, I've always been interested in materials and how you handle them. It's just part of the job as far as I'm concerned.

I think the same willingness on your behalf to research and go beyond established ways of doing things is clear in your redesign of the aluminium seating at Schiphol Airport. Could you tell me more about the process?

I found out through one of my clients that the airport needed to replace their seating and I remember thinking 'Bingo'. Schiphol has such a bad reputation, being a part of the polluting aviation industry, so I was sure that they'd be willing to go the extra mile with a project like this.

Detail of Blink seating designed by Richard Hutten and produced by Lensvelt for Schiphol Airport. The aluminium came from recycled old seating at the airport.

One key driver for the project was that Schiphol Airport had publicly announced that they intend to become a waste-free airport by 2030. This made me look at the airport like a mine that could be mined for raw materials. The aluminium in the existing seats was an obvious source of material for me. Compared with virgin aluminium, recycled aluminium uses only a fraction of the energy needed. When I started digging a little deeper, I realized that all the suppliers I needed to make new seats could be found within a 75-kilometre (roughly 50-mile) radius of the airport, at which point the pieces of the puzzle really started to come together.

Another thing I like about aluminium is that it can be polished without the need for additional coatings or chemicals. I even designed the feet to be made of aluminium because I wanted to keep the construction as free as possible from other materials, such as plastic.

This ties in completely with Schiphol's ambition to become waste-free. The recycled aluminium that you used this time will be there the next time the furniture needs to be replaced – ready to be recycled again without complex material combinations, assemblies or coatings.

Absolutely! I follow the principles of the circular economy and it's my ambition for the product to live on as long as possible. The airport is a rough environment. I found out that it's not so much wear and tear from passengers, rather it's the people working in the airport who drive around in little carts at night, racing each other and bumping into the furniture while cleaning and delivering suitcases and things like that. So I made the furniture easy to disassemble using only screws for repairs and replacing parts.

This also had an impact in terms of cost. Sure, there are less expensive alternatives out there to the furniture that I designed, but seen from the perspective that my design might last for 40 or 50 years, that makes it less expensive than furniture that has to be replaced constantly because it lacks durability. I designed the furniture to have this level of longevity and I hope that for as long as I'm alive it will be there whenever I use the airport.

The scope of this project is very ambitious and potentially quite intimidating to someone who's just starting their journey to becoming a better designer in terms of sustainability. What's your advice for those who may feel a little overwhelmed by this complex topic?

The most important advice I have for young designers is to design products in such a way that, at the end of life, it's possible to take them apart and separate the materials for recycling. I think it's important to be strict about what you mean by sustainable design. Take a closer look at recycling, for example. A lot of people talk about 'upcycling', which as far as I am concerned doesn't exist – it's just a marketing term. Downcycling, however, happens a lot. I don't think that taking, say, plastic packaging and turning it into a park bench is a viable form of recycling. We need to aim higher. Recycling should mean keeping the material properties intact, rather than creating a new material with worse performance, which is not solving anything, only creating new problems.

Blink seating, Schiphol Airport.

Metal recycling

In terms of recycling, the big advantage of metals over many other materials is the relative ease of separating them from other waste. In the case of steel and other magnetic metals, powerful magnets can be used to pull these materials out of mixed waste streams, while a machine called an eddy current separator uses an electric current to separate aluminium and other non-magnetic metals.

Once these metals have been separated from other waste, another round of sorting is required to separate different grades and alloys. Steel is fairly flexible in this respect – steel waste can go into the production of stainless steel, for example – while aluminium recycling is typically more complex. A wide variety of aluminium alloys have been developed to make the material suitable for certain forming processes or to give it certain properties. These alloys have to be sorted and recycled separately to make more of the same alloy. With some aluminium alloys, the requirements for material purity are so high that closed-loop recycling might be the only realistic option. At the other end of the spectrum, so-called foundry alloys allow for more flexibility during the recycling process, but are typically suitable for casting only and cannot be anodized.

As with recycling in general, it's important to distinguish between post-industrial recycled (PIR) and post-consumer recycled (PCR) metals, especially with aluminium waste because the material is so energy-intensive to produce in the first place. Post-industrial scrap should of course always be recycled, but an excess of post-industrial waste implies ineffective processing upstream in the manufacturing process, which should be addressed separately. Some metal suppliers can be vague about whether they use PIR or PCR waste in their recycled materials, so always ask them to confirm the ratio and origin of recycled content.

Coatings and other finishes, and contamination from other materials like plastics that are often used in combination with metals, generally don't have a negative impact on the metal recycling process – the temperatures used when the material is re-melted are typically so high that any coatings, adhesives and other materials simply burn off. Assemblies with parts that are made of metal in combination with other materials should be designed in such a way that they can be separated, using mechanical fasteners or integrated features like click joints and slots.

Global metal recycling rates

The ratio of recycled metals in global production, in million tonnes. Although the average recycling rate for metals is between 60 and 70% in some markets, such as the EU and the US, demand for metals far exceeds the availability of recycled material in other markets, including China, lowering global averages.[1]

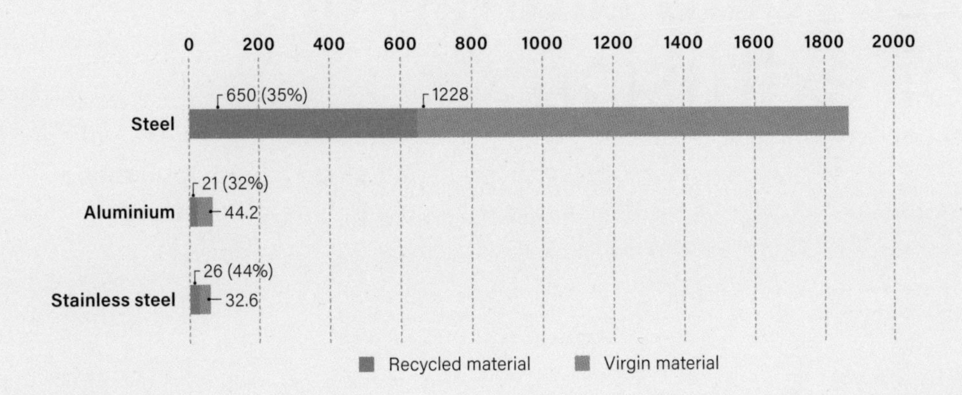

Metal waste streams

Steel and stainless steel can be more straightforward to recycle than aluminium, due to the large number of aluminium alloys in common use. These alloys typically need to be recycled separately, as detailed below.

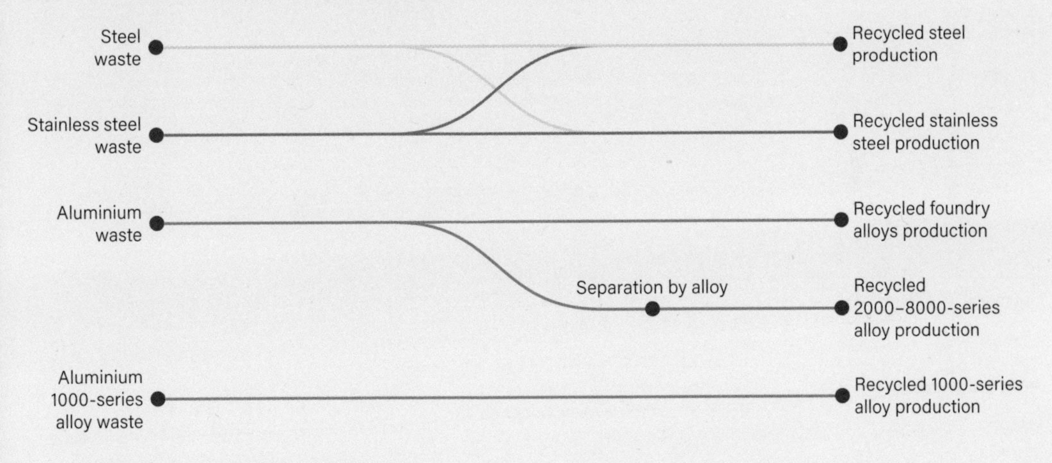

Common aluminium alloys

Key properties and recycling options.

- **1000-series**
  At least 99% pure aluminium – very sensitive to contamination and typically requiring closed-loop recycling. Suitable for forming processes that require high flow, such as impact extrusion.

- **2000- to 8000-series**
  Between 85 and 95% pure aluminium, alloyed with other metals and minerals to provide improved properties and suitability for specific forming and finishing processes, such as sheet forming, extrusion and anodizing. These alloys must be separated during recycling for properties to be retained.

- **4000-series (aka foundry alloys)**
  Alloys that contain silicon, which makes the material easy to cast, but unsuitable for anodizing. These alloys can allow for greater flexibility during the recycling process in terms of mixing different alloys.

Recycled aluminium for extrusion

Specialist suppliers are able to offer 6000-series aluminium alloys with a high ratio of post-consumer recycled content that are suitable for high-quality extrusion applications.

Suppliers & materials	- Hydro CIRCAL recycled 6000-series aluminium - E-MAX X-ECO recycled 6000-series aluminium	
Raw material origin	Extruded profiles from the construction, automotive and consumer product waste. Hydro CIRCAL recycled 6000-series aluminium consists of 75% PCR content and virgin material, while other suppliers typically use a mix of PIR, PCR and virgin material. Ask suppliers to provide ISCC PLUS, SCS Recycled Content or other certification that verifies the ratio of recycled content in specific alloys.	
	Hydro CIRCAL[2]	**Virgin 6000-series aluminium**[3]
GWP	<2.3 kg CO_2e / kg	11.3 kg CO_2e / kg
Energy use	33.2 MJ / kg	163 MJ / kg
Water use	60 l / kg	1,060 l / kg
Toxicity	Aluminium surfaces are considered non-toxic and safe for skin, food and water contact, including in cookware. However, some aluminium finishes may contain toxic ingredients. Ask suppliers to provide a material safety data sheet and confirm the suitability of specific finishes for your application.	
Circularity	Aluminium is currently one of the most widely recycled materials globally and is infinitely recyclable without loss of quality in theory, but different aluminium alloys must be separated during the recycling process to preserve material properties. For an overview of recycling rates and circular design guidelines for aluminium, see page 121.	
Mechanical properties	6000-series aluminium offers a good strength-to-weight ratio, at about one third of the density of steel. It's a rather soft metal, though, so polished, high-gloss surfaces are prone to scratching.	
Environmental resistance	Good heat and corrosion resistance, but raw, unfinished surfaces will react with some chemicals and acids. Good weatherability.	
Forming	Suitable for extrusion, which can be combined with secondary forming such as machining, bending and hydroforming for more complex shapes.	
Finishing	Compatible with many aluminium finishing processes, including powder coating, polishing, sandblasting, anodizing, chemical etching and laser marking. Small imperfections resulting from the recycling process may be visible in the part surface when anodizing with light colours.	

FOLK bicycle stand, designed by Front for Vestre. The extruded profiles are made with Hydro CIRCAL 75% PCR aluminium, with a GWP of about a fifth of the global average for virgin aluminium.

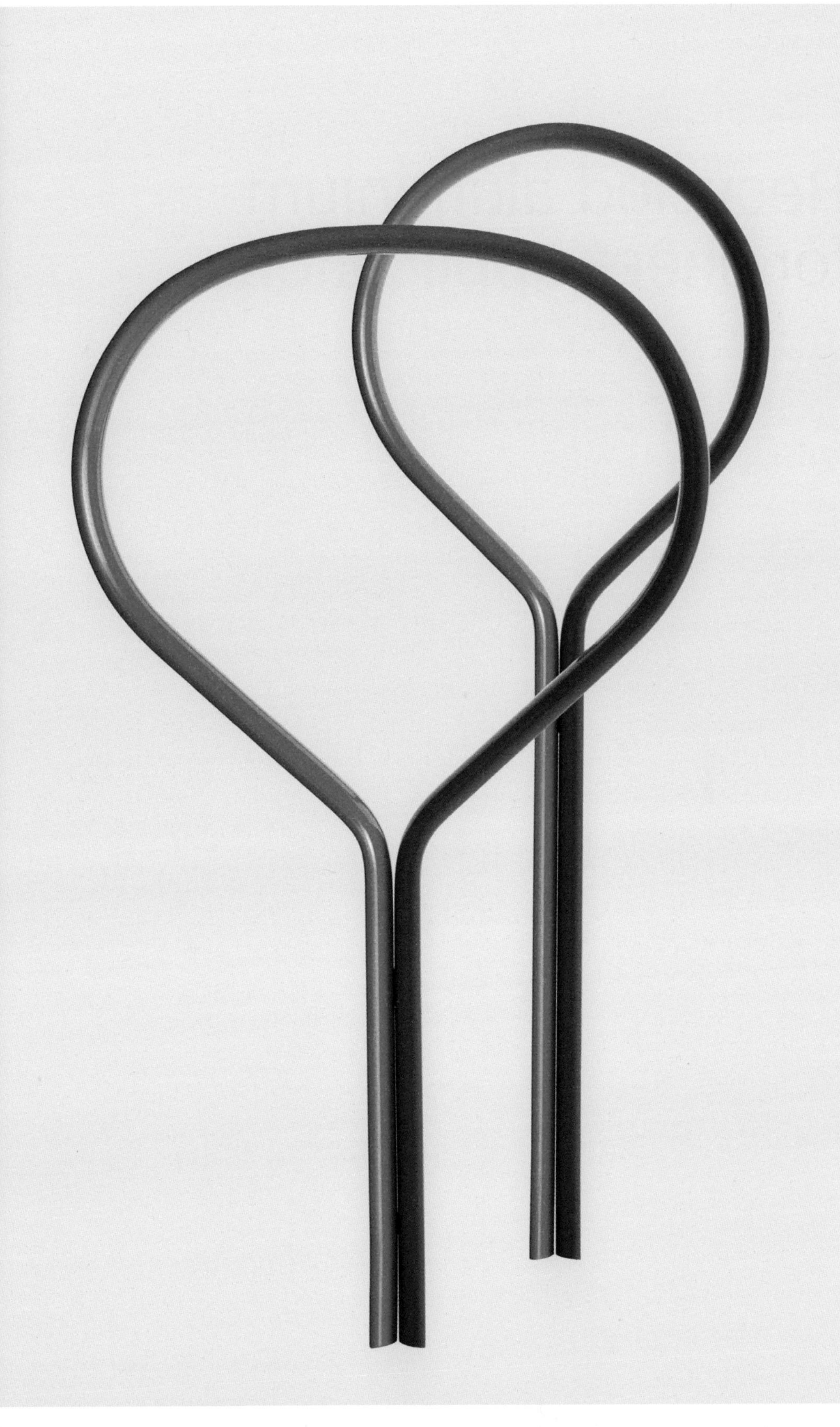

Recycled aluminium for sheet applications

Packaging made with aluminium sheet and foil often uses 3000- and 5000-series aluminium, meaning that these alloys are some of the most widely used materials. This also means that they are some of the most widely recycled aluminium alloys, with many potential applications besides packaging.

Suppliers & materials	- Novelis AL:sust™ recycled 3000- and 5000-series aluminium - Alcoa Ecodura recycled 3000- and 5000-series aluminium	
Raw material origin	Mainly aluminium packaging, such as cans. AL:sust™ aluminium from Novelis contains at least 80% recycled content, while the proportion for Alcoa Ecodura is 50%. Ask suppliers to provide ISCC PLUS, SCS Recycled Content or other certification that verifies the ratio of recycled content in specific alloys.	
	Novelis AL:sust™[4]	**Virgin 6000-series aluminium**[5]
GWP	2 kg CO_2e / kg	11.4 kg CO_2e / kg
Energy use	8.2 MJ / kg	164 MJ / kg
Water use	No data	1,070 l / kg
Toxicity	Aluminium surfaces are considered non-toxic and safe for skin, food and water contact, including in cookware. However, some aluminium finishes may contain toxic ingredients. Ask suppliers to provide material safety documentation and confirm the suitability of specific finishes for your application.	
Circularity	Aluminium is currently one of the most widely recycled materials globally, and 3000- and 5000-series alloys are among the most widely recycled aluminium alloys. Aluminium is infinitely recyclable without loss of quality in theory, but different aluminium alloys must be separated during the recycling process to preserve material properties. For an overview of recycling rates and circular design guidelines for aluminium, see page 121.	
Mechanical properties	3000- and 5000-series aluminium alloys have moderate to high strength and can compete with steel in many applications. It is a rather soft metal, though, so polished, high-gloss surfaces are prone to scratching.	
Environmental resistance	Good heat and corrosion resistance, but raw, unfinished surfaces will react with some chemicals and acids. Good weatherability.	
Forming	3000- and 5000-series aluminium sheet materials are suitable for stamping, pressing and sheet hydroforming, as well as other common sheet metal processes, such as machining and bending.	
Finishing	Compatible with a wide range of common aluminium finishing processes, including powder coating, polishing and anodizing. Sheet can be pre-finished before forming in many cases.	

9000-series LED TV by Samsung. The aluminium exterior of the stand and frame are made with partially recycled aluminium sheet from Novelis.

SAMSUNG

Recycled aluminium for casting

There is a bewildering array of aluminium alloys available for casting – commonly known as foundry alloys – with several suppliers offering materials that contain a high ratio of recycled content.

Suppliers & materials	- Raffmetal SILVAL recycled aluminium foundry alloys - Kuusakoski recycled aluminium foundry alloys - Stena Aluminium recycled aluminium foundry alloys	
Raw material origin	A very wide range of waste from construction, industrial and consumer applications. Recycled foundry alloys are typically made with a mix of PIR, PCR and virgin material. Ask suppliers to provide ISCC PLUS, SCS Recycled Content or other certification that verifies the ratio of recycled content in specific alloys.	
	Raffmetal SILVAL[6]	**Virgin foundry alloys**[7]
GWP	1.9 kg CO_2e / kg	11.3 kg CO_2e / kg
Energy use	30.4 MJ / kg	161.9 MJ / kg
Water use	7.7 l / kg	1,058 l / kg
Toxicity	Aluminium surfaces are considered non-toxic and safe for skin, food and water contact, including in cookware. However, some aluminium finishes may contain toxic ingredients. Ask suppliers to provide material safety documentation and confirm the suitability of specific finishes for your application.	
Circularity	Aluminium is currently one of the most widely recycled materials globally and is infinitely recyclable without loss of quality in theory, but different aluminium alloys must be separated during the recycling process to preserve material properties. For an overview of recycling rates and circular design guidelines for aluminium, see page 121.	
Mechanical properties	Aluminium castings offer good strength-to-weight ratio, at about one third of the density of steel.	
Environmental resistance	Good temperature and corrosion resistance, but raw, unfinished surfaces will react with some chemicals and acids. Good weatherability.	
Forming	Compatible with most common metal-casting processes, including die casting, sand casting and hot isostatic pressing, which can be further formed by machining. Aluminium castings are suitable for large, complex parts like automotive wheel rims, machine parts and furniture.	
Finishing	Compatible with many aluminium finishing processes, including powder coating, polishing, sandblasting, chemical etching and laser marking. Foundry alloys typically contain silicone, which makes them unsuitable for anodizing, often giving visible imperfections in the surface.	

Atlas chair, designed by Johannes Foersom and Peter Hiort-Lorenzen for Lammhults. The base is made with recycled cast aluminium, which has a considerably reduced environmental impact compared with virgin material.

Recycled steel

While the steel industry is responsible for a large proportion of global CO_2 emissions, steel is also one of the most widely recycled materials today, with a growing number of suppliers offering steel materials with a high ratio of recycled content.

Suppliers & materials	- ThyssenKrupp Bluemint® Recycled steel - SSAB EcoSmart™ recycled steel	
Raw material origin	A very wide range of waste from construction, industrial and consumer applications. ThyssenKrupp Bluemint® Recycled is entirely made with a mix of PIR and PCR steel waste, but other recycled steel suppliers typically use a mix of PIR, PCR and virgin material. Ask suppliers to provide ISCC PLUS, SCS Recycled Content or other certification that verifies the ratio of recycled content in specific materials.	
	ThyssenKrupp Bluemint® Recycled[8]	**Virgin steel**[9]
GWP	>0.75 kg CO_2e / kg	2.4 kg CO_2e / kg
Energy use	No data	25.4 MJ / kg
Water use	No data	No data
Toxicity	Steel is considered non-toxic, but some finishes use potentially harmful ingredients. Traditional chrome coatings use hexavalent chromium, a very toxic heavy metal that's safe in finished products, but poses a very real risk to factory workers unless managed properly. More recent chroming processes use trivalent chromium, which is a much safer alternative. Other treatments, such as galvanization, may give off toxic fumes when heated. Ask suppliers to provide material safety documentation and confirm the suitability of specific finishes for your application.	
Circularity	Steel is one of the most widely recycled materials today, partly because it's relatively easy to separate from other waste, but also for the ease of making high-quality recycled materials with steel waste. For an overview of recycling rates and circular design guidelines for steel, see page 121.	
Mechanical properties	Steel offers high strength and surface hardness, but it's a heavy material with a density that's about three times higher than aluminium.	
Environmental resistance	Good temperature resistance, but untreated steel surfaces will corrode and react with chemicals and acids.	
Forming	Steel sheets can be stamped, punched and folded, while rods and billets are suitable for forging. Sheets and tubes can also be hydroformed.	
Finishing	Steel is rarely used without surface treatments that protect the material from corrosion, such as chroming, galvanizing, physical vapour deposition (PVD) or powder coating. Weathering steel (also called corten steel) is an exception, where the steel material is alloyed with copper to give it an oxidized surface. Decorative finishes include polishing, brushing, sandblasting, etching and laser marking.	

S-1500 Chair, designed by Snøhetta for NCP. The legs and support structure are made with recycled steel.

Recycled stainless steel

Like common steel, stainless steel is widely recycled, with several suppliers offering materials with a high recycled content ratio.

Suppliers & materials	- Outokumpu stainless steel - Aperam stainless steel	
Raw material origin	A very wide range of waste from construction, industrial and consumer applications. Stainless steel from the Finnish supplier Outokumpu is made with at least 90% PIR and PCR waste, mixed with virgin material. Ask suppliers to provide ISCC PLUS, SCS Recycled Content or other certification that verifies the ratio of recycled content in specific materials.	
	Outokumpu stainless steel[10]	**Virgin stainless steel**[11]
GWP	3.4 kg CO_2e / kg	7.7 kg CO_2e / kg
Energy use	64.7 MJ / kg	25.4 MJ / kg
Water use	38,200 l / kg	2,350 l / kg
Toxicity	Stainless steel is considered non-toxic and safe for food, water and skin contact. However, a study published in 2013 by researchers at Oregon State University in the US found that stainless steel cookware reacts with acidic food and releases small amounts of nickel and chromium, which could pose a risk for users with allergies. Ask suppliers to provide material safety documentation and confirm the suitability of specific materials for your application.	
Circularity	Stainless steel is widely recycled, and common steel waste can also go into stainless steel production. For an overview of recycling rates and circular design guidelines for stainless steel, see page 121.	
Mechanical properties	Stainless steel offers high strength and surface hardness, but it's a heavy material with a density that's about three times higher than aluminium.	
Environmental resistance	Excellent temperature and corrosion resistance, without the need for secondary coatings.	
Forming	Stainless steel sheets can be stamped, punched and folded, while rods and billets are suitable for forging. Sheets and tubes can also be hydroformed.	
Finishing	Common decorative finishes include physical vapour deposition (PVD), polishing, brushing, sandblasting, etching and laser marking.	

ReNew Scissors by Fiskars, with recycled stainless steel blades and a recycled plastic and cellulose-fibre composite handle.

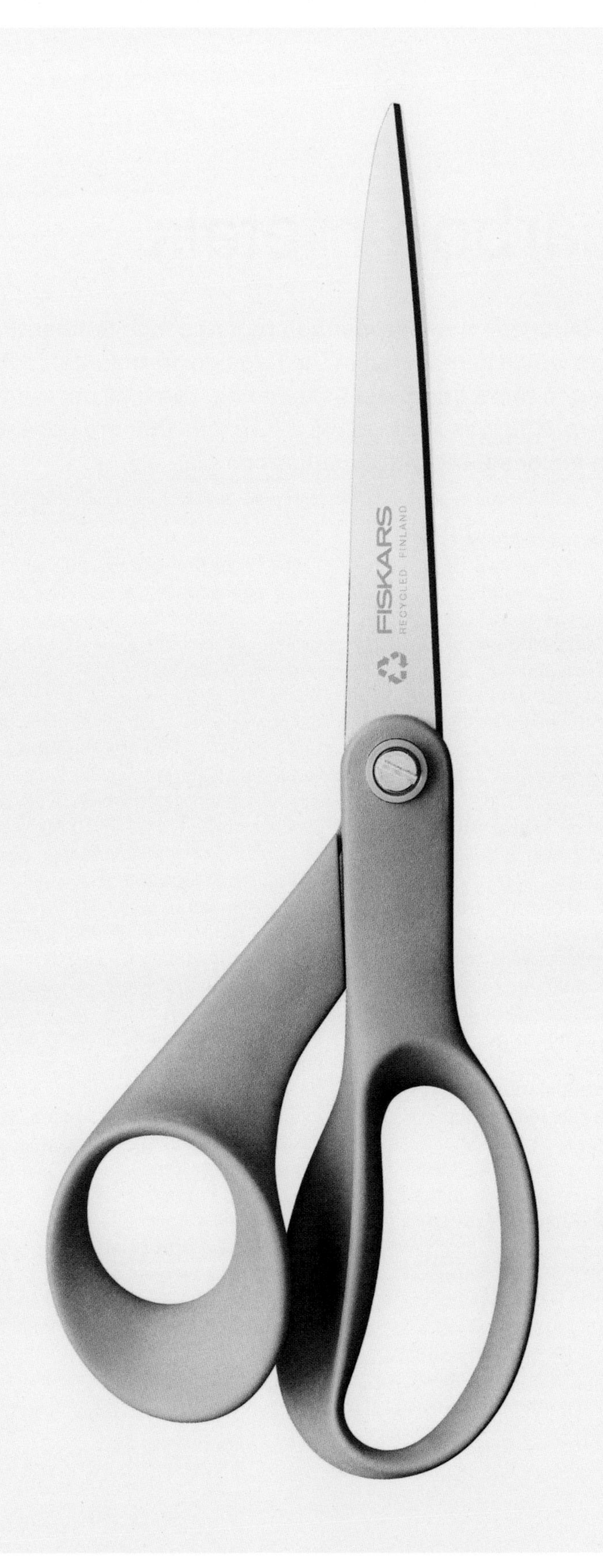
FISKARS
RECYCLED FINLAND

Low-carbon metals

As previously mentioned, metal production requires high temperatures and heavy equipment, which is reflected in the large environmental footprint of metals. According to the World Steel Association, some 72 per cent of global steel production in 2018 was made in blast furnaces that are powered by burning coke – a major source of CO_2 emissions globally.

Aluminium production is even more energy-intensive than steel, and as a result aluminium smelters are often built near sources of renewable energy, such as hydro- and geothermal power plants. However, according to an analysis by the independent energy think-tank Ember, based on data provided by the International Aluminium Institute, more than half of global aluminium production uses coal and natural gas power.

In light of new legislation, such as the European Green Deal that is targeting energy-intensive industries, metal suppliers are looking at alternative sources of energy and ways to improve the efficiency of manufacturing processes. Some steel suppliers, such as ThyssenKrupp in Germany, are replacing traditional iron ore with so-called **direct reduced iron (DRI)**, or **hot briquetted iron (HBI)**, which uses less energy. And in the long term, hydrogen looks set to replace coke as the main source of energy in steel production. The Green Steel Tracker by the Leadership Group for Industry Transition (LeadIT) lists a wide range of decarbonization initiatives in the steel industry. Some of these have already been launched, such as HYBRIT low-carbon steel by the Swedish supplier SSAB (see opposite, and pages 136–37). The HYBRIT process replaces coke and other fossil fuels in the steelmaking process with renewable energy and 'green' hydrogen, which is hydrogen derived from water using renewable energy.

SSAB delivered the first batch of HYBRIT steel to the Volvo Group in late 2021 and several other steel suppliers are developing, or have launched, low-carbon hydrogen-based steel production, including Voestalpine in Europe and HBIS in China.

Compared with steel, the aluminium industry has a longer history of using renewable energy, but there's also potential for improving the overall efficiency of the aluminium production process itself. Hydra REDUXA low-carbon aluminium from the Norwegian supplier Hydro has a carbon footprint of less than 4 kg of CO_2 per kg of material, which Hydro is looking to further improve through efficiency gains in the most energy-intensive steps of the process: alumina refining and electrolysis (see opposite).

Other suppliers are developing new processes altogether, such as ELYSIS, a joint venture between two of the largest aluminium suppliers in the world, Rio Tinto and Alcoa. By replacing conventional electrolysis equipment in aluminium smelters with new technology, the ELYSIS process is capable of removing all GHG emissions that are a direct result of electrolysis. A pilot plant is under construction in Canada and expected to be operational in 2024, but already in 2021 the company delivered small batches of material from its R&D centre to customers in the automotive and consumer electronics industries.

HYBRIT fossil-free steel production

A handful of steelmakers are replacing traditional coal-powered furnaces with a new generation of steel mills that run on hydrogen and renewable electricity. The diagrams below give an overview of the Swedish steel supplier SSAB's HYBRIT process for making fossil-free, zero emissions steel.

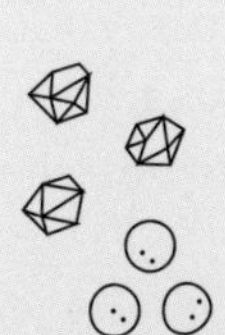

1. Iron ore to pellets
Pellets give improved yield rates and reduced energy consumption in later steps. The process is powered by renewable oil.

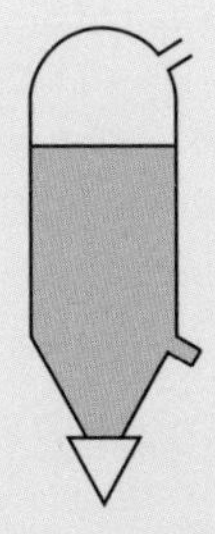

2. Direct reduction of iron
The process is hydrogen-based and uses less energy than coal-powered furnaces, as well as emitting water rather than CO_2.

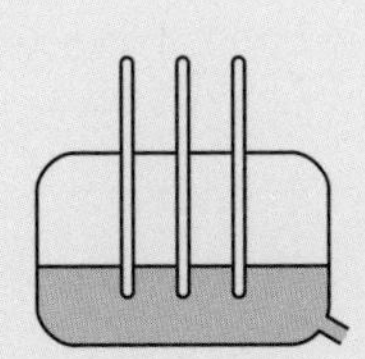

3. Steel alloying
Carbon is added to the iron to produce steel in a furnace that's powered by renewable energy.

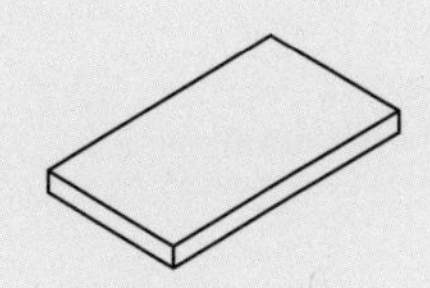

4. Crude steel slabs
The molten steel is formed into slabs for further processing into sheets, rods, wire and other forms of steel.

Hydro REDUXA low-carbon aluminium production

The Norwegian aluminium supplier Hydro has developed a process for making virgin aluminium using renewable energy and improved, energy-efficient processes with 4 kg CO_2e emissions per kg of material produced – at the time of writing, among the lowest emissions of any primary aluminium production in the world.

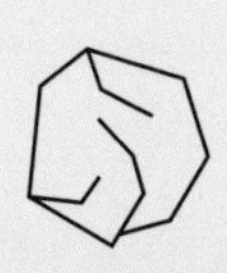

1. Bauxite mining
Hydro uses bauxite from its own mines in Brazil. Hydro's mining operations are certified by the Aluminium Stewardship Initiative (ASI).

2. Alumina refinement
Bauxite is then refined into alumina. Along with electrolysis, this is the most energy-intensive step of the Hydro REDUXA process.

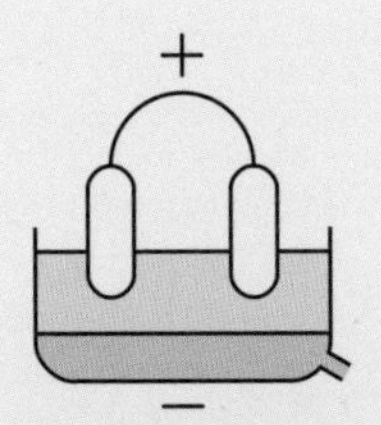

3. Electrolysis
Hydropower-based alumina smelting in an electrolytic cell. Positively charged anodes cause molten aluminium to form at the bottom of the cell.

4. Aluminium ingots
The molten aluminium is further alloyed and cast into ingots for further processing into cast parts, sheet metal and extrusions.

Low-carbon aluminium

Several suppliers offer low-carbon aluminium through the use of renewable energy and efficiency gains in the most energy-intensive steps of the aluminium production process.

Suppliers & materials	- Hydro REDUXA low-carbon aluminium - Alcoa Sustana EcoLum™ low-carbon aluminium - Rio Tinto RenewAl™ low-carbon aluminium	
Raw material origin	Virgin bauxite mining. Ask suppliers to confirm that raw materials extraction complies with ASI, IRMA or other industry standards. Some materials, like RenewAl™ from Rio Tinto, contain a mix of virgin low-carbon aluminium and recycled scrap, in which case ISCC PLUS, SCS Recycled Content or other certification that verifies the ratio of recycled content should be requested.	
	Hydro REDUXA[1]	**Virgin 6000-series aluminium**[2]
GWP	4 kg CO_2e / kg	11.3 kg CO_2e / kg
Energy use	90.6 MJ / kg	163 MJ / kg
Water use	224 l / kg	1,060 l / kg
Toxicity	Aluminium surfaces are considered non-toxic and safe for skin, food and water contact, including in cookware. However, some aluminium finishes may contain toxic ingredients. Ask suppliers to provide material safety documentation and confirm the suitability of specific finishes for your application.	
Circularity	Aluminium is currently one of the most widely recycled materials globally and is infinitely recyclable without loss of quality in theory, but different aluminium alloys must be separated during the recycling process to preserve material properties. For an overview of recycling rates and circular design guidelines for aluminium, see page 121.	
Mechanical properties	Good strength-to-weight ratio, at about one third of the density of steel. Aluminium is a rather soft metal, though, so polished, high-gloss surfaces are prone to scratching.	
Environmental resistance	Good heat and corrosion resistance, but raw, unfinished surfaces will react with some chemicals and acids. Good weatherability.	
Forming	Available in grades suitable for extrusion, casting, sheet and wire applications. Aluminium parts can be further refined with machining, folding and hydroforming for more complex shapes.	
Finishing	Compatible with many aluminium finishing processes, including powder coating, polishing, sandblasting, anodizing, chemical etching and laser marking.	

NXT90 ERGO stroller by Emmaljunga. The chassis is made with Hydro REDUXA low-carbon aluminium profiles assembled with screws rather than rivets, allowing for interchangeable parts and repairs.

Emmaljunga

Low-carbon steel

The handful of low-carbon steel materials that are currently available commercially replace fossil fuels with green hydrogen and renewable energy, giving a significantly reduced environmental footprint compared with conventional steelmaking processes.

Suppliers & materials	- ThyssenKrupp Bluemint® Pure low-carbon steel - SSAB HYBRIT low-carbon steel - Voestalpine Greentec low-carbon steel	
Raw material origin	Virgin iron ore. Ask suppliers to confirm that virgin iron-ore mining complies with ResponsibleSteel™, IRMA or other relevant standards and certification.	
	ThyssenKrupp Bluemint® Pure[3]	**Virgin steel**[4]
GWP	0.6 kg CO_2e / kg	2.4 kg CO_2e / kg
Energy use	No data	25.4 MJ / kg
Water use	No data	No data
Toxicity	Untreated steel is considered non-toxic, but some steel finishes use potentially harmful ingredients. Conventional chrome coatings use hexavalent chromium, a very toxic heavy metal that, although safe in finished products, is a risk to factory workers unless managed properly. More recent chroming processes use trivalent chromium, which is a safer alternative. Galvanized steel, another common finish, gives off zinc oxide fumes when heated, so is not suitable for cooking utensils or food contact in general. Ask suppliers to provide a material safety data sheet and confirm the suitability of specific finishes for your application.	
Circularity	Steel is one of the most widely recycled materials today, partly because it's relatively easy to separate from other waste, but also due to the ease of making high-quality recycled materials with steel waste. For an overview of recycling rates and circular design guidelines for steel, see page 121.	
Mechanical properties	Steel offers high strength and surface hardness, but it's a heavy material with a density that's about three times higher than aluminium.	
Environmental resistance	Good temperature resistance, but untreated steel surfaces will corrode and react with chemicals and acids.	
Forming	Steel rods and billets are suitable for forging, while sheets can be stamped, punched and folded. Steel sheets and tubes can also be hydroformed.	
Finishing	Steel is rarely used without surface treatments that protect the material from corrosion, such as chroming, galvanizing, physical vapour deposition (PVD) or powder coating. Weathering steel (or corten steel) is an exception, being alloyed with copper to give an oxidized surface. Decorative finishes include polishing, brushing, sandblasting, etching and laser marking.	

Candle holder, designed by Lena Bergström for the Swedish steelmaker SSAB, made using the world's very first fossil-free steel. The steel is made with iron produced using their HYBRIT technology, reduced by using 100% hydrogen instead of coal and coke (see page 133).

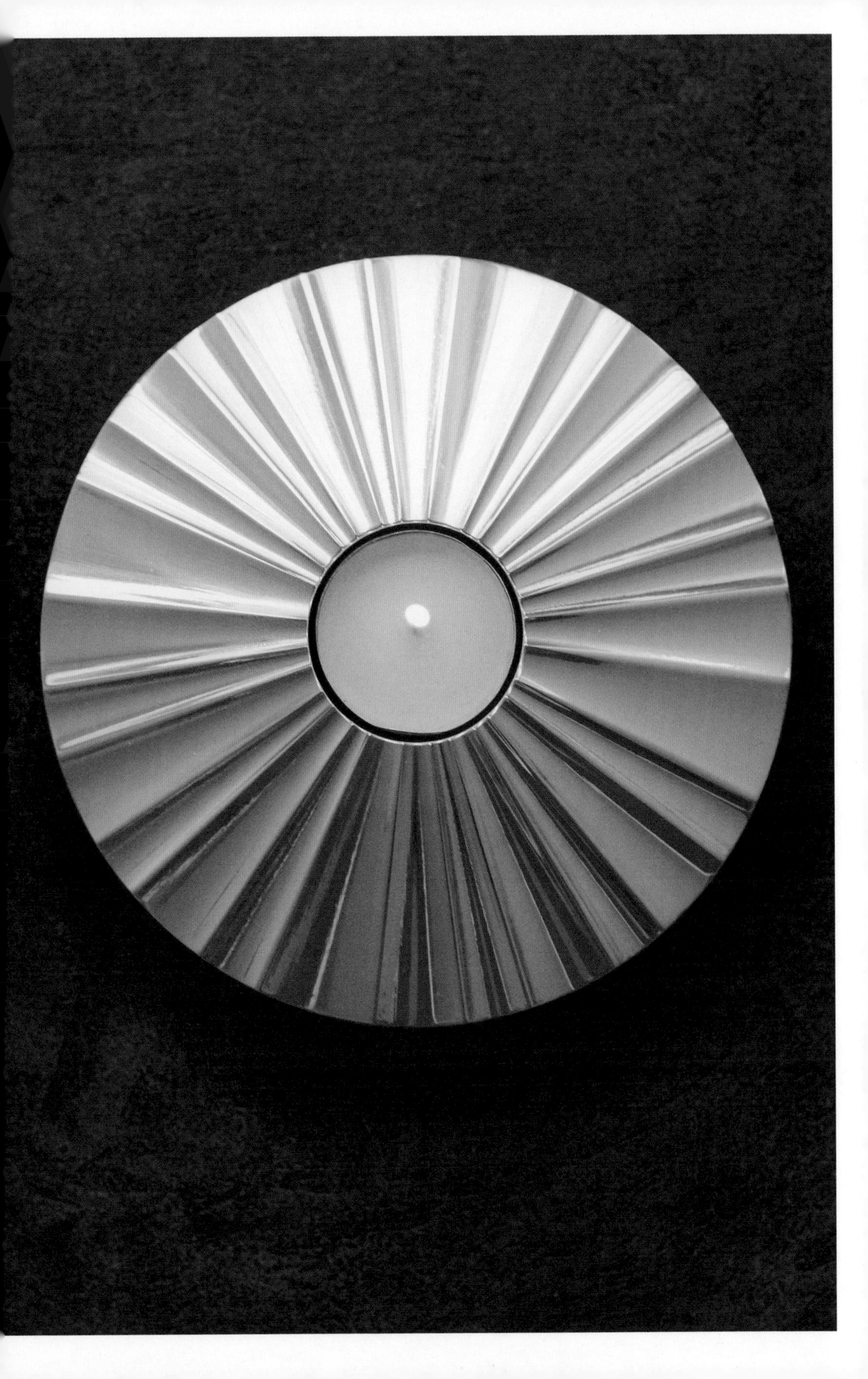

4
Ceramics and glass

Ceramics and glass are often grouped together because they share the same origin: minerals. Most ceramics are made of clay, while glass materials are made of dry mineral powder (mainly silicon dioxide from sand). Despite being some of the most abundant materials on the planet, these minerals are non-renewable and must be mined. Large-scale mining leaves permanent scars on the land and, unless mines are carefully managed, minerals and by-products can leak out, contaminating soil and water. Ceramic and glass raw materials are now also running out in some areas with a long history of mining, leading to illegal mining in some cases. Illegal sand mining is of particular concern, endangering shorelines and river deltas around the world. Several certifications exist for the responsible extraction and production of ceramic materials, including the Initiative for Responsible Mining Assurance (IRMA).

Beyond mining, processing also contributes to the environmental impact of ceramics and glass, requiring high temperatures and a lot of energy. Several projects aim to develop more efficient processes (many can be found in the Inventory of European Research and Innovation Projects published by Cerame-Unie, the European Ceramic Industry Association), but currently, most ceramics and glass materials are made using traditional electric- or gas-powered kilns and glass furnaces.

Ceramics and glass are generally inert and non-toxic, but coatings and decorative finishes need special attention. Ceramics typically require a protective glaze to withstand moisture and chemicals, but these are also used to add colour, gloss and other decorative effects. In the past, glazes typically used toxic materials like lead, cadmium and chromium, which can withstand high temperatures, but several non-toxic alternatives are now available. For the same reason, decorative glass finishes once often used lead and other heavy metals.

End-of-life options vary greatly. While some materials, such as packaging glass, are widely recycled, options are slimmer for other types of glass and ceramics. As mentioned in the interview with Ward Massa on the following pages, very large volumes of ceramic and glass waste are generated by the construction industry, representing over one third of all annual waste in the EU, for example. On the other hand, ceramic and glass are extremely durable if properly looked after.

Origins of ceramic materials

Ceramic and glass materials are made with minerals that exist in different forms in nature, from huge solid blocks of stone to fine sand and clay.

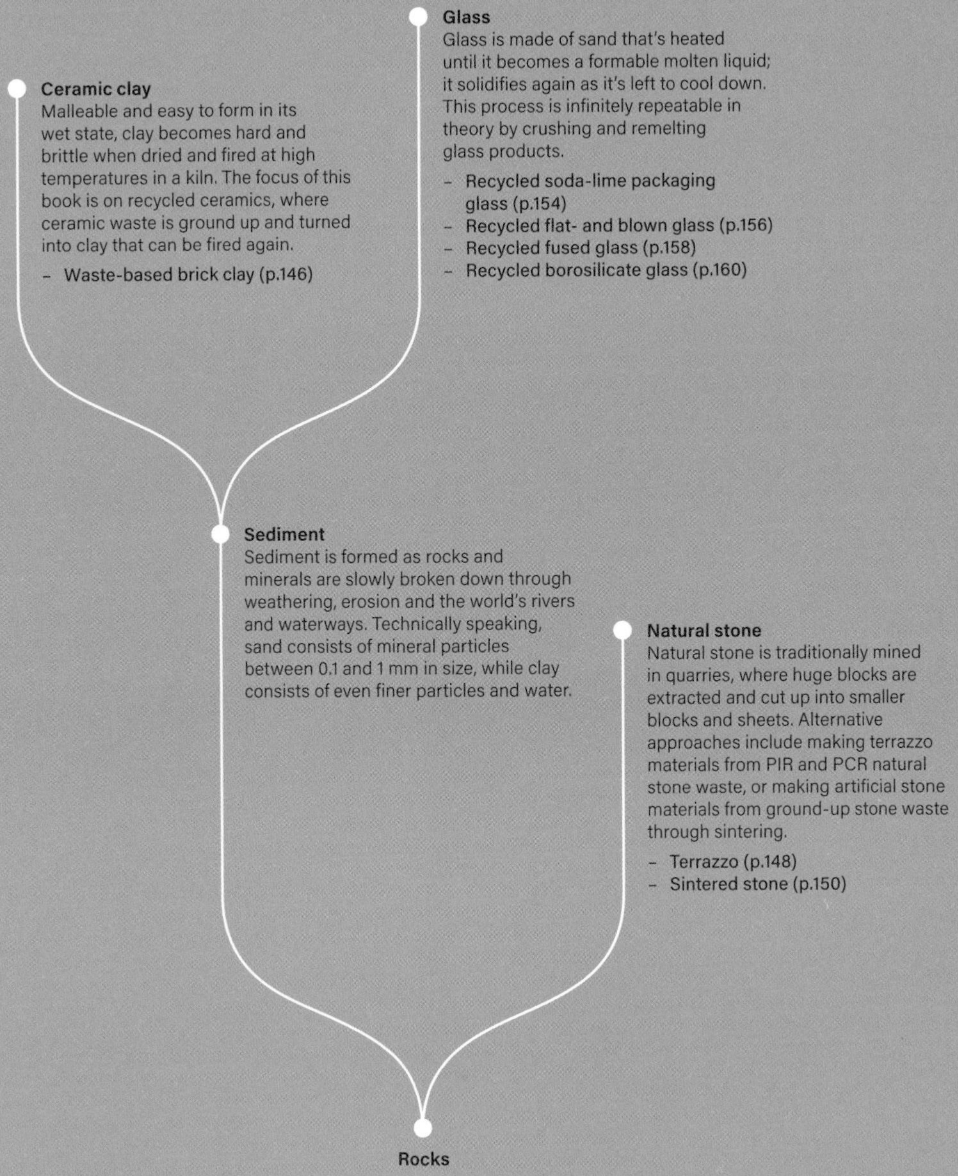

Glass

Glass is made of sand that's heated until it becomes a formable molten liquid; it solidifies again as it's left to cool down. This process is infinitely repeatable in theory by crushing and remelting glass products.

- Recycled soda-lime packaging glass (p.154)
- Recycled flat- and blown glass (p.156)
- Recycled fused glass (p.158)
- Recycled borosilicate glass (p.160)

Ceramic clay

Malleable and easy to form in its wet state, clay becomes hard and brittle when dried and fired at high temperatures in a kiln. The focus of this book is on recycled ceramics, where ceramic waste is ground up and turned into clay that can be fired again.

- Waste-based brick clay (p.146)

Sediment

Sediment is formed as rocks and minerals are slowly broken down through weathering, erosion and the world's rivers and waterways. Technically speaking, sand consists of mineral particles between 0.1 and 1 mm in size, while clay consists of even finer particles and water.

Natural stone

Natural stone is traditionally mined in quarries, where huge blocks are extracted and cut up into smaller blocks and sheets. Alternative approaches include making terrazzo materials from PIR and PCR natural stone waste, or making artificial stone materials from ground-up stone waste through sintering.

- Terrazzo (p.148)
- Sintered stone (p.150)

Rocks

'We want to show that it's possible to make something beautiful from waste.'

An interview with Ward Massa, co-founder and director of StoneCycling, an Amsterdam-based start-up that makes bricks and tiles from various ceramic waste streams. Find out more at stonecycling.com

As an outsider, it seems to me that StoneCycling has come a long way in a short time. Could you tell me a bit about how you started the company and where you are today?

I started StoneCycling with my friend Tom van Soest in 2011. Tom is a designer and invented the StoneCycling process while he was a student at the Design Academy in Eindhoven, while my background is in communication and politics. I've always been interested in systems and how systems change, and I knew that StoneCycling had the potential to make a positive impact in the construction industry. But there is, of course, a huge difference between having a good idea and scaling it up to a point where it becomes a viable product. We've been building a network of manufacturing partners, demolition companies and recyclers that take in certain waste streams that we can use. Sometimes we use waste sourced at a demolition site that goes into a new building on the same site, such as the House for the City in Helmond in the Netherlands. Our first big project was for a residential house in Rotterdam in 2016, but it's only in the last couple of years that the business has really started to grow, which is connected to new legislation and incentives in the construction industry in the European Union and elsewhere that have made our products interesting to a much wider group of potential customers. In 2022, the first building in the US to use our bricks was finished.

Ceramic construction waste
at the StoneCycling factory.

Sustainability is complex and it can be difficult to decide what to prioritize, but it seems that you had a clear idea from the start about where StoneCycling could make a difference and have the biggest impact.

Tom and I found out quite early on that anywhere you go on the planet, the construction industry is the single biggest source of waste – in Europe alone we're talking about some 850 million tonnes of waste from demolition work and building sites, as well as industrial waste from material suppliers, which amounts to about a third of all the waste generated in the EU. We were quite surprised to find out that this waste stream is so big; I don't think it's a well-known fact at all. Also, while the amount of waste is growing, we could see that virgin construction materials are running out, which is leading to some really depressing things like illegal sand mining, which is in turn emerging as a serious threat to beaches and rivers across the world. So in this context we had an idea about a basic recipe for making new ceramic products from what is essentially the biggest waste stream on the planet. Why not make bricks so that some of this waste can be put back into the construction industry while preserving virgin materials at the same time?

One of the most striking things about StoneCycling is how beautiful the bricks are. Did you have a strong aesthetic vision for where you wanted to go with the material from the start?

We always wanted to show that it's possible to make something beautiful from waste. I think we also knew that making the material as beautiful as possible would improve our chances of making it into a mainstream material, rather than a niche product. Now we're seeing more established companies in the building industry adopting some of the things that we do at StoneCycling, which is great because it confirms the appeal of our products, and together we can have a bigger impact. I do think, though, that because we're not a traditional material supplier, we're able to be more flexible and clear about what we do, so really we should be ahead of the rest of the industry, moving faster and trying new things when it comes to sustainability and aesthetics.

On that note, how do you think StoneCycling will evolve in the future?

At this point we're in a position where we're able to offer bricks that are made with waste, which is great, but they still have to be fired in a kiln, which uses lots of energy. Currently, the best we can do is using natural gas that we compensate for through a carbon-offsetting scheme, which we know is not good enough, by far. So we're looking into alternative ways of firing the bricks, such as kilns that are powered by hydrogen.

Ultimately, though, I feel that our ability to bring about change on the energy side of things is limited. We're in a much better position to develop the raw materials aspect, and the kind of waste streams that we're able to use – not to mention what happens to our own bricks when the time comes for them to be recycled. We also have a very exciting collaboration with the US start-up Biomason, which has developed a process for growing ceramic materials at low temperatures with the help of the same microorganisms that create coral reefs in nature.

WasteBasedBrick® materials by StoneCycling (see also pages 146–47).

Ceramic recycling

Circularity, modularity and other jargon often used to describe cutting-edge sustainable material technology seems firmly embedded in the DNA of ceramics. After all, a well-preserved brick from a demolition site could be picked up and put right back into a new building. And yet, ceramic recycling doesn't happen anywhere near as much as it could.

There are two basic approaches to ceramic recycling today – recycling of post-industrial mining waste materials and recycling of post-industrial manufacturing waste and post-consumer ceramic products. Waste materials from the mining industry are often contaminated with heavy metals and other impurities that are difficult and expensive to remove, but new rules and legislation mean that some of this waste is becoming a viable source of raw material. For example, by-products from metal smelting and bauxite mining have been explored as raw materials for ceramic products by the UK-based design consultancy Studio ThusThat. Other waste streams, such as dust from natural stone cutting and grinding, can be used for making clay for new ceramic products, or **sintered** and made into solid stone surfaces.

Recycling of manufacturing waste and post-consumer ceramic products is more complex, as it requires either smashing up ceramic parts into chunks for recycling in terrazzo-type materials, or grinding them down into a fine powder for making clay. Some suppliers, such as StoneCycling, are able to take in both industrial waste and post-consumer ceramic products as raw materials for recycled ceramics production.

Currently, however, most post-consumer ceramic waste does not go back into the production of new ceramic materials. According to a report by the European Environment Agency published in 2020, recycled materials only represent about 3 to 4 per cent of all materials used in new buildings.[1] According to the same report, most recovered construction materials are either landfilled, downcycled and used as a base material in road construction, for example, or 'backfilled', which means that the material is used as an alternative to excavated earth in noise barriers and other landscaping projects

As mentioned already, ceramics forming and finishing is quite energy-intensive, and this applies to recycled materials too. So the main environmental benefit of recycled ceramics today is perhaps found in reducing the amount of virgin materials used, as well as the amount of waste generated by the ceramics industry, rather than reducing emissions, although this could change in the future as more energy-efficient ceramic processes become available.

Ceramic waste and recycling

Annual construction waste in the EU, in million tonnes, of which ceramic waste materials make up a large proportion. Currently, most ceramic waste is either downcycled and used in road construction and other infrastructure projects, or backfilled (see opposite). The remaining waste is landfilled.[2]

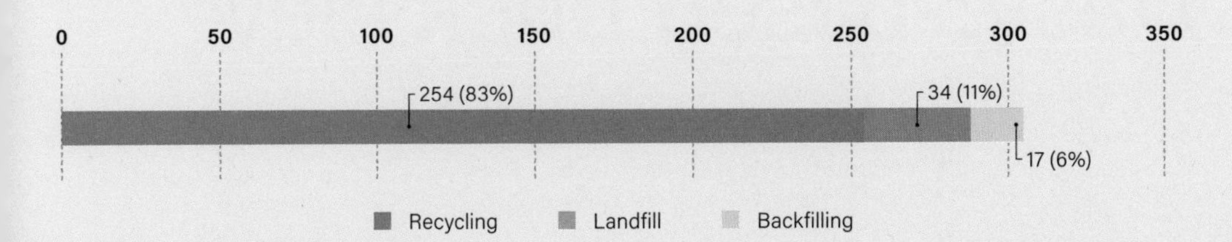

Ceramic waste streams

Common ceramic manufacturing processes, such as cutting and grinding, result in large volumes of ceramic powder that can be used in recycled ceramics. Post-industrial and post-consumer ceramic product waste can be ground up or crushed to make recycled clay or composites like terrazzo materials.

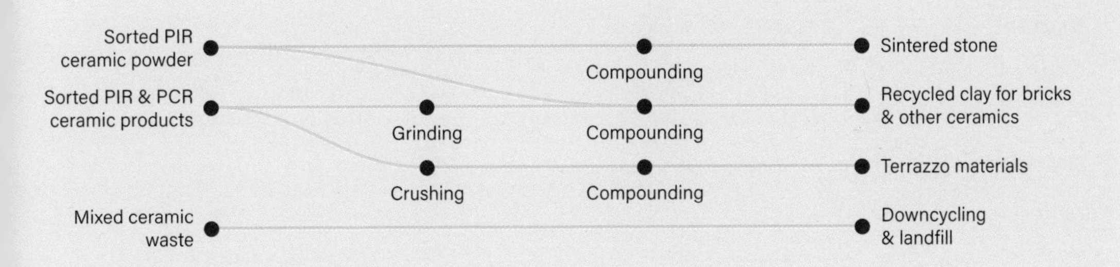

The StoneCycling process

The Dutch start-up StoneCycling has developed a process for recycling various types of industrial ceramic waste, such as dust from natural stone cutting, but also post-consumer ceramic construction materials and sanitary ware. For more on StoneCycling, see pages 146–47, as well as the interview with co-founder Ward Massa on pages 140–43.

Waste collection
Sources include post-industrial and demolition waste. The ceramic waste is then sorted by type.

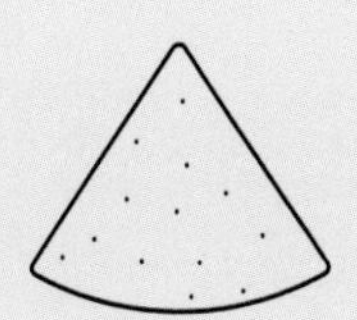

Grinding & compounding
The waste is ground into a fine powder, mixed with other ingredients and turned into clay with colour and texture.

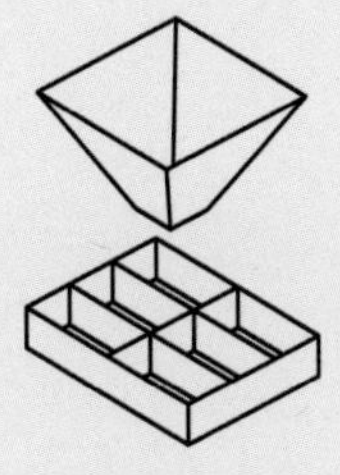

Shaping & firing
The recycled clay is poured into a compression mould and fired in a kiln. Custom shapes can be developed.

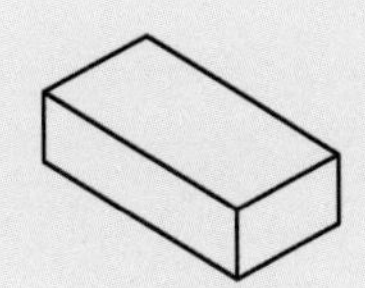

WasteBasedBricks®
Recycled bricks from StoneCycling are available in a wide range of standard colours and finishes.

Waste-based brick clay

Waste-based brick clay made with ground-up raw materials sourced from various post-industrial and post-consumer waste streams.

Suppliers & materials	- StoneCycling WasteBasedBricks® - Kenoteq K-BRIQ	
Raw material origin	PCR waste bricks, but also other ceramic waste, such as sanitary ware and PIR waste from ceramic material manufacturing. The ratio of waste material used typically varies between 60 and 100%; ask suppliers to confirm the origin and ratio of recycled content.	
	StoneCycling WasteBasedBricks®[3]	**Virgin bricks**[4]
GWP	0.2 kg CO_2e / kg	0.7 kg CO_2e / kg
Energy use	No data	1.6 MJ / kg
Water use	No data	12 l / kg
Toxicity	StoneCycling's WasteBasedBricks® undergo the same tests as virgin bricks according to EU safety standards. Confirm the safety of the ingredients used in glazed bricks, as well as general material safety documentation and relevant certification for the intended application.	
Circularity	Bricks from StoneCycling and Kenoteq can either be reused or ground up and recycled into new bricks. For an overview of recycling rates and circular design guidelines for ceramics, see page 145.	
Mechanical properties	Very durable, with high compression strength, although somewhat brittle on impact.	
Environmental resistance	Very good thermal, UV and moisture resistance. Good resistance to most chemicals.	
Forming	Bricks and other shapes made with brick clay are first formed in a mould, then dried and fired at high temperature. Bricks and other moulded shapes can be further formed by machining, water-jet cutting and other ceramic cutting processes.	
Finishing	Both StoneCycling and Kenoteq offer a range of standard colours and finishes, as well as a bespoke service for large orders. StoneCycling offer several finishes in terms of smoothness and gloss level. Further finishing options can be achieved using polishing, sandblasting, etching and other ceramic finishing processes.	

WasteBasedBricks® by StoneCycling. StoneCycling is able to process a wide range of ceramic waste, including PIR factory waste such as ceramic dust from cutting, as well as PCR construction material and sanitary wares.

Terrazzo

Originally invented as a way to recycle natural stone offcuts in Italy in the 1600s, terrazzo materials include different composite materials made with ceramic and glass waste materials and a variety of binders.

Suppliers & materials	- Herrljunga Terrazzo - Alusid SILICASTONE™ Terrazzo - Altrock terrazzo
Raw material origin	Natural stone offcuts, glass waste and other aggregate that can be cast and polished together with a binder. While cement and plastic resin are common binders in terrazzo materials, Alusid SILICASTONE™ uses recycled glass as a binder. Ask suppliers to confirm the type of binder and ratio of recycled content in specific materials.
	Herrljunga Terrazzo, HT-LYKKE[5]
GWP	0.2 kg CO_2e / kg
Energy use	2.1 MJ / kg
Water use	0.4 l / kg
Toxicity	Terrazzo materials that use cement as the matrix material are considered inert and non-toxic, although this may vary with materials that use other binders. Ask suppliers to provide material safety documentation and relevant certification for the intended application.
Circularity	Specialist recyclers can recycle terrazzo sheets into other products, such as tiles. For an overview of recycling rates and circular design guidelines for ceramics, see page 145.
Mechanical properties	Very durable with high compression strength, although the material is somewhat brittle on impact. Large terrazzo-covered areas may crack, as the material tends to expand and contract in response to changes in temperature and humidity. This can be solved by dividing the material into smaller sections using divider strips.
Environmental resistance	Terrazzo materials that use cement as the matrix material have good thermal, UV and moisture resistance, as well as good resistance to most chemicals, but the properties of other binders may vary.
Forming	Terrazzo materials are supplied in sheets and blocks that can be further formed by machining, water-jet cutting and other ceramic cutting processes.
Finishing	Terrazzo materials offer a wide range of aesthetic options, from the size, shape, colour and type of aggregate to the colour and smoothness of the concrete matrix. Terrazzo sheets and blocks can be further finished using polishing, sandblasting, etching and other ceramic-finishing processes.

Io Wall Light, designed by Hand&Eye Studio. The lampshade is made with SILICASTONE™ from Alusid, a composite material made with glass and ceramic waste with a minimum recycled content of 98%.

Sintered stone

While sintering processes are energy-intensive, recycled raw materials can be used in the production of sintered stone materials, reducing ceramic waste and the use of non-renewable natural stone materials.

Suppliers & materials	- Cosentino Dekton sintered stone - Neolith sintered stone - IDYLIUM sintered stone	
Raw material origin	Fine powder from various ceramic waste streams, mixed with virgin material. Ask suppliers to confirm the exact ratio for specific materials.	
	Cosentino Dekton sintered stone[6]	**Virgin marble**[7]
GWP	0.7 kg CO_2e / kg	0.3 kg CO_2e / kg
Energy use	9.7 MJ / kg	5.2 MJ / kg
Water use	161 l / kg	2.3 l / kg
Toxicity	Sintered stone materials consist entirely of minerals and are considered inert and non-toxic. While these materials are typically approved for food and water contact, ask suppliers to confirm the suitability for specific applications.	
Circularity	Sintered stone materials can either be reused or ground up and recycled into new material. For an overview of recycling rates and circular design guidelines for ceramics, see page 145.	
Mechanical properties	Very durable, with high compression strength, although somewhat brittle on impact.	
Environmental resistance	Very good thermal, UV and moisture resistance. Good resistance to most chemicals. Confirm the suitability of the material for specific applications with suppliers.	
Forming	Supplied in sheets and blocks that can be further formed by machining, water-jet cutting and other ceramic cutting processes.	
Finishing	The aesthetics of sintered stone materials can be customized to a high degree by selecting the type of powder, pigments and composition as the material is being made. Most suppliers offer a range of standard colours and textures, with bespoke options available for large orders. Sintered stone sheets and blocks can be further finished using polishing, sandblasting, etching and other ceramic finishing processes.	

Talyd Vase, designed by Aljoud Lootah for the Dekton Capsule Collection by Cosentino. Dekton is a sintered stone material that can be partially made with recycled material.

Glass recycling

Glass recycling is much more established than ceramics recycling, but rates are still fairly low compared with materials like metals and paper. According to the recycling trade magazine *Recovery*, only about 21 per cent of the total amount of glass produced globally in 2018 was recycled, although recycling rates are higher in individual markets like the EU and US.

In terms of glass materials, soda-lime glass is used for bottles, jars and other packaging, as well as tableware, window glass and more. It accounts for about 90 per cent of global production, with most of the remainder made up of borosilicate glass, which has good thermal shock resistance and is often found in cookware, medical applications and car headlamps. Other types of glass include glass ceramics, which have even better thermal shock resistance and are mostly used in stovetops and industrial applications, and aluminosilicate glass, which is much more impact and scratch resistant than other glass materials and is common in display covers in mobile devices.

Currently, nearly all recycled glass is soda-lime glass. Products fall into two categories – packaging glass and everything else, including flat glass used in windows, mirrors, furniture and cars, as well as blown-glass products like tableware and accessories. According to the European Federation of Glass Recyclers (FERVER), 9.8 million tonnes of soda-lime glass were recycled in the EU in 2018 – 8.3 million tonnes were packaging glass; the remainder was flat glass. While there are a few recyclers that specialize in borosilicate glass, recycling of glass ceramics and aluminosilicate glass is rare.

As with other materials, recycled glass comes with the risk of contamination. Window, car and other non-packaging glass is often laminated or coated, which can discolour and otherwise contaminate recycled glass materials unless these finishes can be separated out. Packaging glass recycling is typically less complicated, although some decorative finishes, such as colourants and enamels, have relied on lead, cobalt, cadmium and other heavy metals in the past. In terms of these substances leaching out from recycled food and liquid packaging glass, a study commissioned by the UK Food Standards Agency back in 2000 concluded that while there's a small risk of migration in scratched and worn recycled glass, as well as recycled glass that comes into contact with certain acids, levels are low and should not discourage recycling. Some markets limit the ratio of substances of concern in recycled packaging to 200 ppm, so confirm what rules apply to your project.

Colour is another factor in glass recycling. Most packaging glass is clear, brown or green, and this is separated before recycling to preserve colour consistency. Otherwise, there are no differences between virgin and recycled glass in terms of aesthetics and performance.

Using recycled glass makes it possible to run glass furnaces at lower temperatures, reducing the energy needed by 2 to 3 per cent for every 10 per cent of recycled glass added to the mix. Recycling also preserves non-renewable virgin raw materials – mainly sand, a commodity that is running out in many parts of the world.

Glass recycling globally and in the EU

Glass recycling statistics tend to focus on packaging glass, while numbers for flat glass used in construction, automotive and many other industries are harder to find. This chart is an estimate of global and EU glass recycling rates in million tonnes at the time of writing.[1]

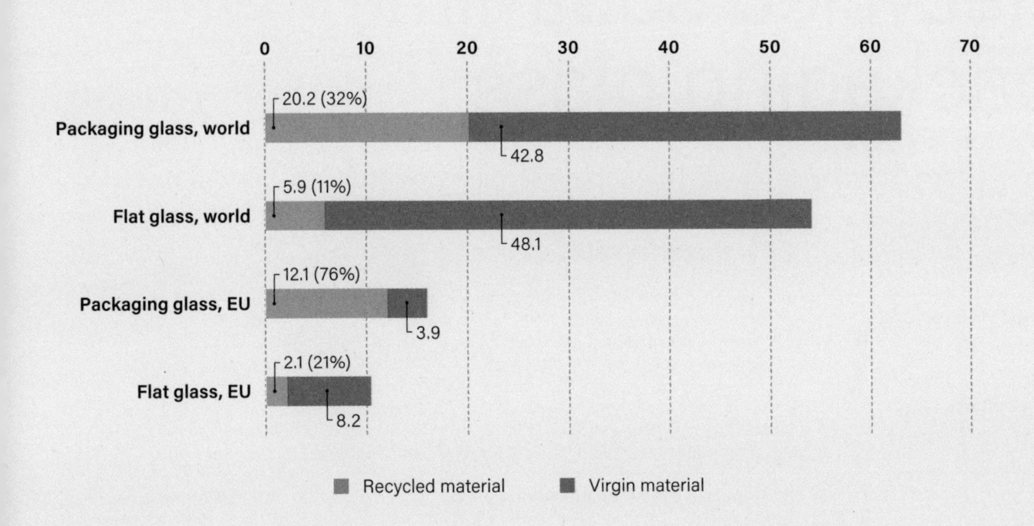

Glass waste streams

While closed-loop recycling of packaging glass is common, glassware and other consumer products are more difficult to recycle, as any film laminates or decorative coatings used will contaminate other glass waste. Unlaminated and uncoated flat glass can be recycled into window glass and other products. Small volumes of borosilicate glass are recycled, while recycling of glass ceramics and aluminosilicate glass is rare.

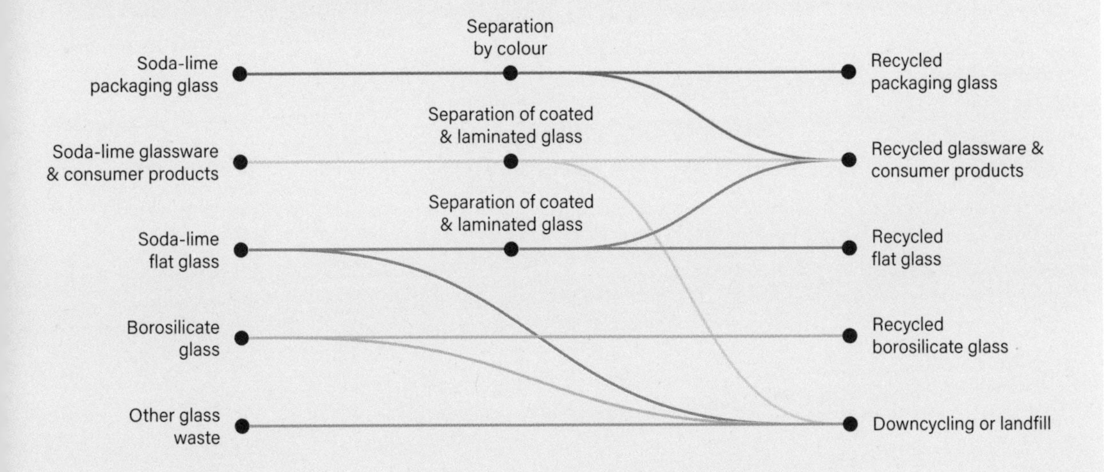

Recycled soda-lime packaging glass

This is by far the most widely recycled type of glass, with many suppliers offering up to 80% recycled glass in bottles and other containers.

Suppliers & materials	- Vetropack recycled soda-lime packaging glass - Enva recycled soda-lime packaging glass - O-I recycled soda-lime packaging glass	
Raw material origin	Bottles, jars and other packaging glass. Recycled packaging glass is likely to contain a combination of PCR and PIR waste, as well as virgin material. Ask suppliers to confirm the ratio of recycled content and origin of recycled raw materials.	
	80% recycled soda-lime packaging glass[2]	**Virgin soda-lime packaging glass**[3]
GWP	1 kg CO_2e / kg	1.1 kg CO_2e / kg
Energy use	13.5 MJ / kg	13.5 MJ / kg
Water use	2,000 l / kg	2,000 l / kg
Toxicity	While glass is considered inert and non-toxic, in the past, some decorative glass finishes and colourants contained lead and other heavy metals. Studies have shown that there's a small risk of migration when recycled packaging glass is scratched and in contact with certain acids, so some markets restrict the ratio of substances of concern in packaging glass to 200 ppm. Ask suppliers to provide material safety documentation and confirm the suitability for food-contact applications.	
Circularity	Soda-lime packaging glass is widely recycled and infinitely recyclable in theory. For an overview of recycling rates and circular design guidelines for glass, see page 153.	
Mechanical properties	Good scratch resistance and compression strength, but poor impact resistance.	
Environmental resistance	Excellent UV and chemical resistance, but the material doesn't withstand thermal shock very well.	
Forming	The vast majority of packaging glass is blow-moulded, but additional compatible forming processes include compression moulding, water-jet cutting and other glass cutting processes.	
Finishing	Recycled packaging glass is typically available in three colours – clear, green and brown. While colour can be added to clear recycled glass during the manufacturing process, the potential for adding colour to green and brown recycled glass is more limited. Packaging glass can be further finished using in-mould textures, sandblasting, etching and other glass finishing processes.	

Echovai bottle by Vetropack. The soda-lime glass in the bottle is thermally hardened, giving it thinner walls and a lower weight than standard glass bottles. Echovai bottles can be recycled with other packaging glass waste.

Recycled flat and blown glass

Flat glass in windows and mirrors, and blown glass tableware, lighting and other interior accessories, is recycled at a lower rate than packaging glass. In Europe, flat glass manufacturers use about 25% recycled glass on average.[4]

Suppliers & materials	- Saint-Gobain DIAMANT soda-lime flat glass with at least 30% recycled content - Le Verre Beldi soda-lime blown glass with up to 100% recycled content - Maltha recycled **glass cullet** from flat and blown soda-lime glass waste	
Raw material origin	Window glass, mirrors, car glass, tableware, lighting and other flat- and blown glass applications. The recycled material ratio is likely to be considerably lower compared with packaging glass; ask suppliers to confirm the exact ratio for specific materials.	
	Saint-Gobain DIAMANT 30% recycled flat glass[5]	**Virgin flat glass**[6]
GWP	1.1 kg CO_2e / kg	1.3 kg CO_2e / kg
Energy use	15.1 MJ / kg	17.9 MJ / kg
Water use	2.6 l / kg	2.3 l / kg
Toxicity	While glass is considered inert and non-toxic, in the past, some decorative glass finishes and colourants contained lead and other heavy metals. Studies have shown that there's a small risk of migration when recycled glass is scratched and in contact with certain acids. Ask suppliers to provide material safety documentation and confirm the suitability for food-contact applications.	
Circularity	Soda-lime glass is widely recycled and infinitely recyclable in theory. For an overview of recycling rates and circular design guidelines for glass, see page 153.	
Mechanical properties	Good scratch resistance and compression strength, but poor impact resistance and limited tensile strength.	
Environmental resistance	Excellent UV and chemical resistance, but the material doesn't withstand thermal shock very well. Applications that require rapid heating and cooling, such as cookware, should use another glass type, such as borosilicate glass or glass ceramics.	
Forming	Compatible forming processes include glass blowing, blow moulding, floating and slumping. Glass sheets and 3D parts can be further formed by water-jet cutting and other glass cutting processes.	
Finishing	The colour of recycled glass waste will have an impact on the ability to add colour to recycled glass – clear glass waste offers most flexibility, but any colour added will make the glass less valuable for recycling at the end of life. Flat and blown glass can be further finished using slumping, in-mould textures, sandblasting, etching and other glass finishing processes.	

Moroccan Vase by HAY, made with hand-blown and painted recycled soda-lime glass.

Recycled fused glass

Fused glass sheets and blocks are made with recycled glass from post-industrial waste, with the crushed pieces of glass cullet clearly visible in the surface.

Suppliers & materials	- MAGNA Glaskeramik®
Raw material origin	Industrial and packaging glass waste. MAGNA Glaskeramik® materials are made with 100% waste.
	MAGNA Glaskeramik®[7]
GWP	1.5 kg CO_2e / kg
Energy use	68.3 MJ / kg
Water use	3,442 l / kg
Toxicity	Fused glass is fully made with recycled glass, without the use of synthetic binders, and is considered inert and non-toxic. Ask the supplier to provide material safety documentation and confirm the suitability for certain applications.
Circularity	Recyclable with other flat soda-lime glass. For an overview of recycling rates and circular design guidelines for glass, see page 153.
Mechanical properties	Good scratch resistance, as well as compression and tensile strength, with decent impact resistance.
Environmental resistance	All-round good environmental resistance, with excellent UV, chemical and stain resistance, and good thermal resistance.
Forming	Supplied in sheets and blocks that can be further formed using machining, water-jet cutting and other glass cutting processes.
Finishing	Fused glass is available in a range of standard colours, with custom colours available for larger orders, either through sourcing coloured waste or by adding pigments to the recycled glass mix during manufacturing. Surface textures can be added by using slumping, sandblasting, etching and other glass finishing processes.

Samples by Magna Glaskeramik®. The material is made with PIR glass waste that is heated up and compressed to form sheets and blocks.

Recycled borosilicate glass

Borosilicate glass (or Pyrex) has much better thermal shock resistance than soda-lime glass, so is suited to cookware and other applications that undergo rapid heating and cooling. Very little borosilicate glass is recycled, by just a handful of specialist suppliers.

Suppliers & materials	- Vitro Minerals recycled borosilicate glass	
Raw material origin	Cook- and labware, medical- and industrial-glass applications. Vitro Minerals offer up to 100% recycled borosilicate glass grades.	
	100% recycled borosilicate glass[8]	**Virgin borosilicate glass**[9]
GWP	1.9 kg CO_2e / kg	2.4 kg CO_2e / kg
Energy use	29.4 MJ / kg	37.9 MJ / kg
Water use	No data	No data
Toxicity	Borosilicate glass is considered inert and non-toxic. Ask suppliers to provide material safety documentation and confirm the suitability for certain applications.	
Circularity	Currently, very little borosilicate glass is recycled, possibly due to much lower volumes of the material being used compared with soda-lime glass. For an overview of recycling rates and circular design guidelines for glass, see page 153.	
Mechanical properties	Good scratch resistance and compression strength, but poor impact resistance.	
Environmental resistance	Excellent chemical, stain and thermal resistance. Borosilicate glass can withstand sudden temperature changes of up to 180°C (350°F).	
Forming	Blow moulding, casting and continuous drawing for tubes, as well as floating for flat sheets. Borosilicate glass can be further formed using machining, water-jet cutting and other glass cutting processes.	
Finishing	Pigments can be added to clear borosilicate glass waste to create coloured recycled material. Further finishing can be achieved using sandblasting, etching and other glass finishing processes.	

Surface Straws by Layer Design, made with recycled borosilicate glass. Borosilicate glass is currently recycled to a much lesser extent than soda-lime glass.

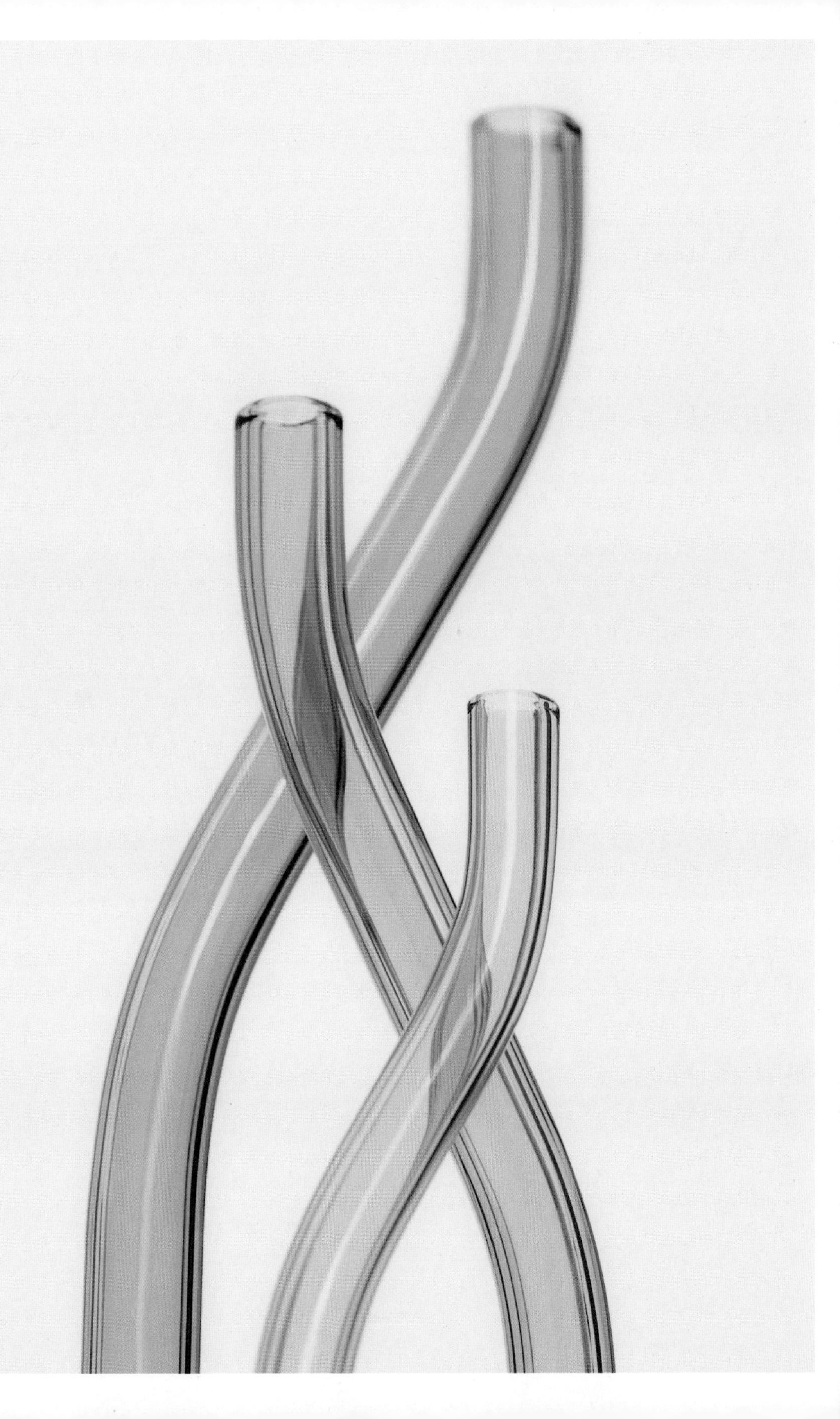

5
Wood

The importance of trees and forests to the global environment cannot be overstated, providing shelter and habitats for other plants and animal life, as well as the cultural and economic backbone of local communities. Trees also have a pivotal role in absorbing CO_2 from the atmosphere within the fast carbon cycle (see page 10), storing it until they are dead or decaying, or burnt in forest fires or as biofuel. About half the volume of a tree consists of water and small quantities of nutrients absorbed from the forest floor, such as nitrogen and sulphur. The other half consists of carbon, literally pulled from thin air. Like other renewable raw materials, trees absorb CO_2 while growing, so some suppliers declare the estimated amount of absorbed ('biogenic') CO_2 stored in wood as a separate item from CO_2 emissions that are the result of material processing. Because wood processing is quite energy efficient, biogenic CO_2 often exceeds emissions from processing, leading to a negative carbon footprint.

Compared with most agricultural materials grown for food and materials production, forests are very efficient, needing only rain- and groundwater to grow. This, added to the relative ease of harvesting, is a large part of the reason why wood has been used as a material and fuel throughout human history. As a result, the effects of deforestation are also all too familiar. Poorly managed forests run the risk of losing the ecosystems that have evolved under the protective cover of their trees. Soil erosion is a particular concern because the sun will dry out the fertile topsoil where a forest has once grown, causing it to be washed away over time and making it difficult, if not impossible, for new growth to become established.

The Forest Stewardship Council (FSC) and the Programme for the Endorsement of Forest Certification (PEFC) both offer certifications for responsible forestry, but some question whether they do enough to protect biodiversity in forests. Different tree species support different animal and plant life, and it's even thought that interactions between different tree species regulate forest temperatures. Other aspects, such as the role of decaying trees, which return nutrients to the soil and provide fertile habitats for fungi, insects and other plant and animal life, should also be taken into account. Always ask suppliers how they approach sustainable forestry and the long-term environmental impact of the materials they produce.

Greenhouse gases absorbed by global forests

The chart below estimates the annual capacity of the world's forests to absorb CO_2e emissions, measured in billion tonnes. This should be seen in light of deforestation – removing forests to make space for agriculture, urban areas and other uses significantly reduces the capacity of forests to absorb emissions.[1]

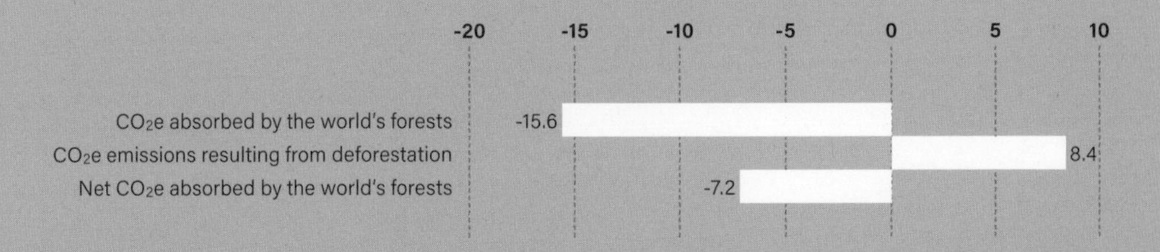

Global certified forest area

According to PEFC and FSC, some 460 million hectares – or just over 11% – of the world's forests were certified by either organization or both in 2020. Both organizations promote preservation and enlargement of forests, as well as biodiversity, workers' rights and pay, and reducing hazardous chemicals in the forest industry.[2]

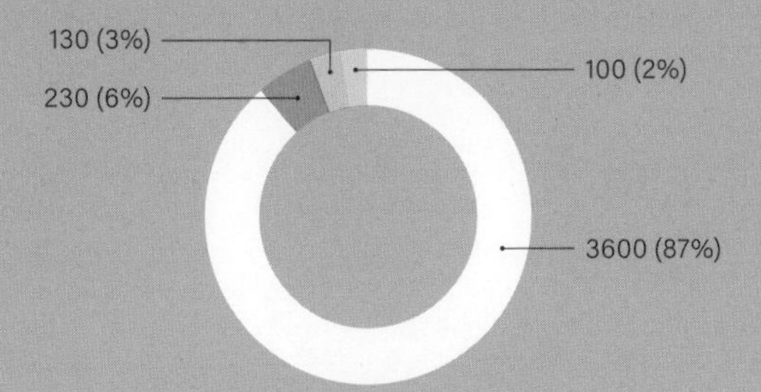

Global timber trade

The majority of the global timber trade is solid wood, at close to a staggering 500 million cubic metres annually, including logs, beams, planks and other standard timber sizes. Although information about the volumes of PEFC- and FSC-certified timber is not readily available, the rough share can be estimated using the chart above.[3]

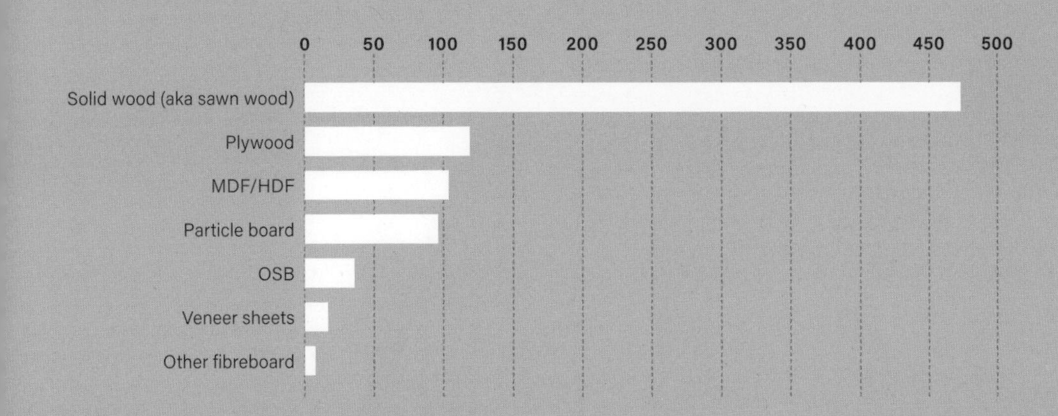

'If someone developed this new material that just grows by itself in forests, you'd think it was some kind of space-age super material.'

An interview with Henrik Taudorf Lorensen, founder and CEO of TAKT, a Danish furniture brand that goes further than most in terms of being open and transparent about its environmental footprint and business model. Find out more at taktcph.com

TAKT stands out in the world of furniture for the level of information that you're willing to share about your business and its impact on the environment. Could you tell me how you arrived at this decision?

It goes back to my experience of launching and leading B&O Play, taking this brand that was 90 years old at the time and a bit exclusive – almost extravagant – and turning it into something relevant and accessible for a younger and much wider audience. When we launched TAKT, I wanted to do something similar for Danish furniture design. Given how expensive what we now regard as Danish design classics are today, it's easy to forget that they were meant to be approachable and accessible to the emerging middle class when they were first designed in the post-war era. But I was also interested in exploring what a genuine take on sustainability would look like for a furniture brand.

Detail of the Soft Chair, designed by Thomas Bentzen for TAKT.

Five years ago, when we started talking about this, it wasn't unusual to see furniture brands that would take an existing product, change the upholstery fabric to a 10 per cent recycled wool material and label it their 'sustainable collection'. This seemed a bit provocative to me. We quickly realized that we wanted to do more, and to be transparent about how we did it. I genuinely believe that younger consumers today are incredibly smart at distinguishing between marketing messages versus how things are really put together, so unless we're able to back up the claims we make, some consumers might think it all sounds too good to be true. We decided we'd be consistent about getting relevant eco-certificates and labels for our entire portfolio, and that we'd publish data on where our environmental impact was coming from. Our website is designed to be like an infinitely knowledgeable salesperson who's on hand to explain everything to customers, from carbon footprint and eco-certificates to production and the real cost of our products. Because we've been serious about sustainability from the start, we can also be open about it. I believe this gives us a competitive advantage – it's almost handing the burden of proof to our competitors and saying, if we can be open about our environmental footprint, why can't you?

Looking at the TAKT portfolio so far, there is little doubt that wood occupies a special place for the brand. On the one hand it's a very traditional material, and on the other it could not be more contemporary in the context of sustainability.

If someone came along and said that they'd developed this new material that just grows by itself in forests, sucking carbon out of the atmosphere during the process, you'd think that it was some kind of space-age super material, wouldn't you? Add to this that wood is simple and efficient to process, has the strength to make very durable products, and has a rare beauty and warm tactility that will continue to evolve over the years, and you have a truly remarkable material. And as if that's not enough, when you discard wood products they will only release the carbon and other ingredients stored inside them back into the biological cycle. In terms of sustainable sourcing of wood, however, we have to take several things into account. First of all, we do not source any wood species that are in any way endangered, and we use certified wood materials only. We are also making small strides in opening the eyes of consumers to other types of wood beyond oak, which tends to be the most sought-after choice for furniture. Up here in the Nordic countries, we have a lot of pine and beech, which are excellent for furniture, but it's a challenge to make furniture that has the same level of attractiveness for consumers using these materials.

It seems that there's a lot of room for exploring wood finishes. What is your thinking around this?

Currently we use either water-based lacquer or oil treatments, which makes sense environmentally, but we'd like to switch to oil-based finishes as much as we can because they're so easy for the user to maintain – any stains can be removed simply by lightly sanding them down and reapplying the oil finish. As a brand, we're driven by making a difference, and as far as I'm concerned there's no reason not to be super-excited about the current wave of new approaches to designing sustainably with wood. A hundred years from now we'll look back on some of these designs as classics.

Cross Chair, designed by Pearson Lloyd for TAKT. The chair consists of four wooden parts that are assembled with four screws, allowing for the chair to be flat-packed and also easily repaired.

Solid wood

From an environmental point of view, perhaps the biggest challenge of solid wood as a material for design lies in the fact that different wood species are valued so differently. Traditional wood applications such as furniture and architecture tend to value some species much more highly than others, which in the worst cases has led to illegal logging of the most sought-after types.

The International Union for Conservation of Nature (IUCN) maintains a red list of wood species that are in danger of extinction. It should also be noted that different wood species grow at different rates - once felled, it can take anywhere between 40 and 150 years for new trees to grow, depending on the specific species, which of course has a major impact on the management and continuity of forests.

The most common certifications for solid wood are FSC and PEFC, but other relevant certificates include the EU Ecolabel, and Timber Origin and Legality (Origine et Légalité des Boise; OLB), issued by Bureau Veritas. However, while these organizations are both well established and widely trusted, their scope and impact have been questioned, leading to an ongoing discussion that can be followed through independent forums and organizations such as fsc-watch.com.

Increasingly, issues such as biodiversity and the ability of forests to support animal and insect life are coming into focus. While it's estimated that forest monocultures - neat rows of identical trees planted solely for maximizing the yield of specific timber types or key raw materials like natural rubber - only account for about 3 per cent of global forests at the time of writing, it's easy to see how larger future plantations might negatively impact local ecosystems and communities.

After logging, timber needs to be dried before further processing. Air drying is the most energy-efficient option; kiln drying can make wood more durable as the higher heat will remove any mould, mildew and insects, although it uses more energy. Once dried, wood is relatively energy-efficient to process compared with many other materials, such as metals and plastics, which require high temperatures and heavy equipment for forming. Additionally, many sawmills are powered by burning by-products such as sawdust and bark.

Solid wood sawing patterns and grain direction

Different sawing patterns will define the grain direction, visual appearance and performance of wood, but also the efficiency and yield of felled timber. Any sawdust and remaining wood in the negative spaces after cutting is typically used to power the sawmill, or as raw material for engineered wood (see page 172).

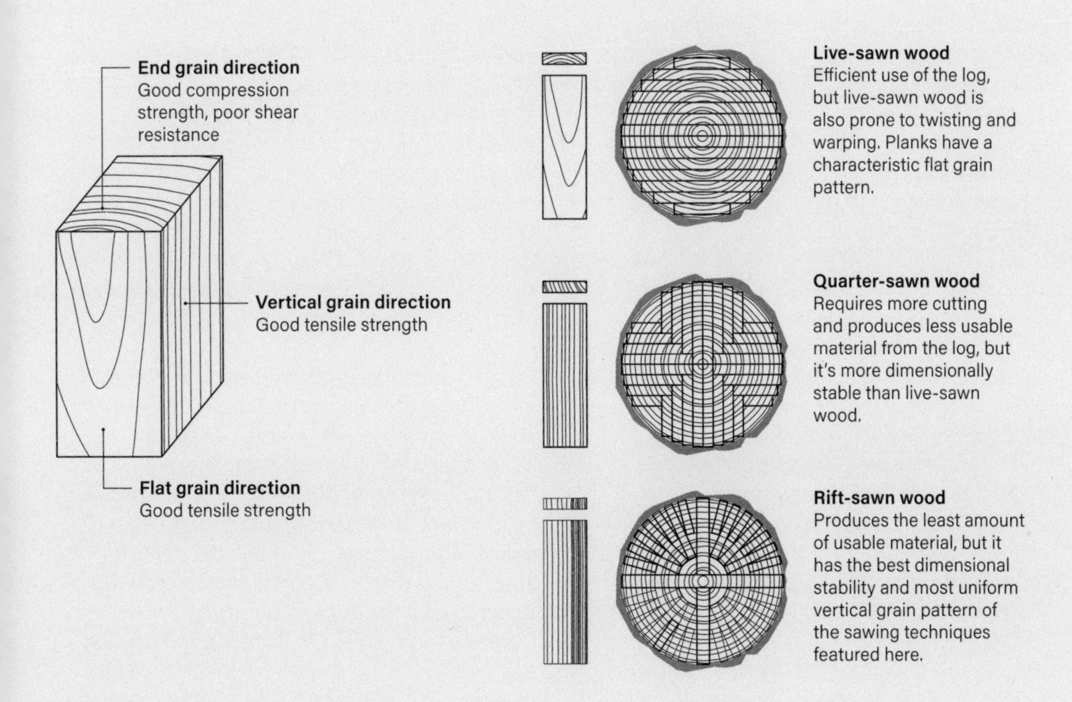

Forest rotation rates

The number of years it takes for different wood species to reach the optimal age for harvesting (with hatched markings indicating a range). More recently, so-called short rotation forestry using fast-growing species is generating a lot of interest for its ability to absorb large amounts of CO_2 in a short amount of time.[4]

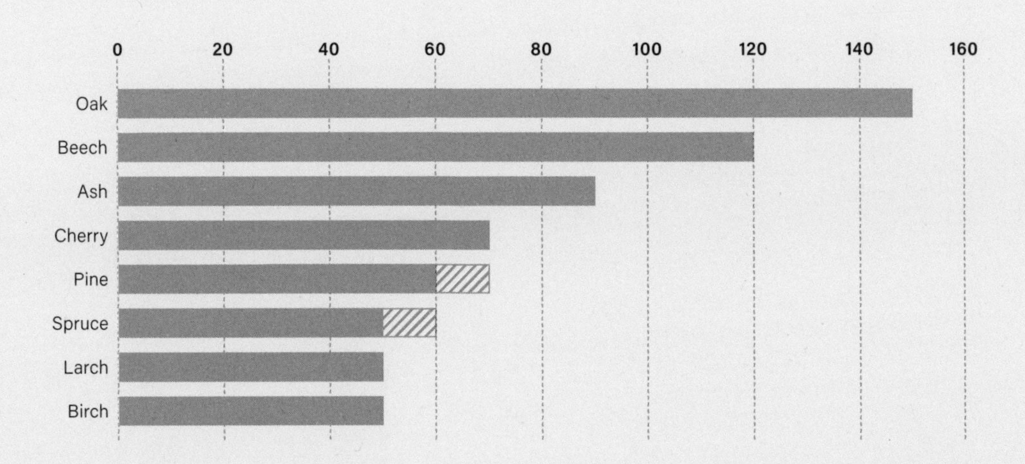

Grain direction (shown on page 169) is another aspect that needs to be taken into account when designing with solid wood (as shown opposite). The tensile strength of wood is considerably higher along the length of the grain, as opposed to parallel to it, while the compression strength is highest in the direction of the end grain.

Solid wood also typically needs some kind of surface treatment to resist moisture, grease and UV radiation, not to mention adding colour, gloss level or other aesthetic properties. While solvent-based wood lacquers and varnishes are common, be aware that these coatings often contain volatile organic compounds (VOCs) that cause emissions during application, and that finished wood products can continue to emit toxic fumes for long periods after the coating has been applied. Water-based wood coatings are an increasingly common alternative that give off fewer or no VOCs at all, depending on the formulations. A wide range of traditional, renewable and non-toxic wood treatments are also available, such as linseed oil and beeswax, as well as mineral-, animal- and plant-based pigments, which can be used for wood staining.

Wood finishes have a major impact on the end-of-life options for wood waste in terms of recycling, composting and use as biofuel. A report published in 2018 by BioReg, an EU-funded project aimed at improving wood recycling rates in Europe, lists several examples of wood recycling from around the EU, such as wood packaging waste – typically pallets – and post-industrial solid wood being shredded and recycled into engineered wood materials or other products altogether, like plant mulch.

Painted and lacquered solid wood is currently not widely recycled, and in many cases not suitable for composting either, as these coatings are typically not biodegradable. This leaves landfilling or burning the wood waste as a biofuel as the only viable options. This paints a rather bleak picture in the context of the wood waste being generated globally. According to the European Furniture Industries Confederation (EFIC), in 2018, some 11 million tonnes of used furniture were thrown out in Europe alone, of which a significant proportion is believed to have been made from wood.

Left to right, from top
Row 1: Scots pine, Norway spruce, European larch
Row 2: aspen, silver birch, alder
Row 3: beech, pedunculate oak, wych elm
Row 4: wild cherry, Norway maple, ash

Engineered wood

Engineered wood materials are typically made with post-industrial waste from solid wood production, such as sawdust, offcuts, and crooked and knotty timber that's unsuitable for solid wood production, but post-consumer recycled engineered wood is becoming more common.

Plywood is one of the earliest engineered wood materials, consisting of layers of glued veneers that can be oriented to provide enhanced tensile strength in a specific direction. Medium-density fibreboard (MDF), a composite of fine wood particles in a resin matrix, has the same mechanical properties in all directions. Other examples, like oriented strand board (OSB) and parallel strand lumber, use flakes or shredded timber to offer enhanced strength.

Although these materials offer enhanced mechanical performance and efficient use of waste materials, they tend to use much more energy in production than solid wood. They are also traditionally made with formaldehyde-based resin binders. Urea-formaldehyde (UF) is of particular concern – mostly to factory workers, and subsequently in the form of dust when engineered wood materials are cut during fabrication, but to a lesser extent to consumers too; products containing UF resin continue to emit low levels of formaldehyde long after they're made. Phenol-formaldehyde (PF) resin is an alternative with lower emissions and, in most places, it's a legal requirement that engineered wood for use indoors uses PF resin. UF resin is still common in outdoor construction applications.

Beyond resins, it's important to know that wood naturally contains small amounts of formaldehyde, meaning that no wood-based product can claim to be entirely free of this toxic organic compound. Several certifications limit the use of formaldehyde-based resins in engineered wood materials, including the California Air Resources Board (CARB) 'no added formaldehyde' (NAF) and 'ultra-low emitting formaldehyde' (ULEF) certificates, as well as the US-based consumer and industrial-safety consultancy UL Solutions' GREENGUARD certificate. See opposite for more on both of these.

As an alternative to formaldehyde, NAF engineered wood materials typically use adhesives made with methylene diphenyl diisocyanate (MDI), isocyanates and polyurethane instead, which are thought to have lower VOC emissions. And a few suppliers have developed new types of binders based on renewable raw materials, such as Evertree's MDF Green and Dekodur's® ECO-HPL.

The vast majority of engineered wood materials are not biodegradable or compostable because the resin binders used are non-degradable. And although most engineered wood materials are recyclable in theory, many recyclers are reluctant to process them because of the health risks of materials that contain formaldehyde. This is likely to change, however, as formaldehyde-based resins are being increasingly phased out and new wood recycling processes are becoming available, such as recycled MDF from the Belgian supplier UNILIN Panels.

Global wood waste volumes

Post-industrial and post-consumer wood waste, measured in million m³. Logging waste, such as knotty and bent wood, as well as small branches, is typically shredded for use in particleboard and OSB. Sawmill waste, including sawdust and other residue, is used in MDF and similar materials. A much smaller amount of post-consumer wood waste is recovered and recycled into raw material for engineered wood products.[5]

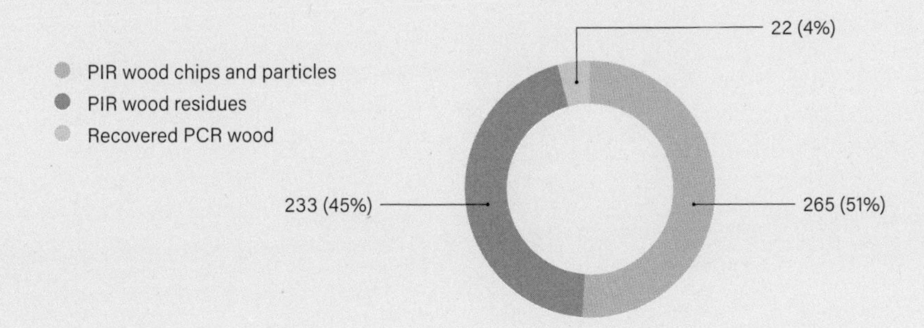

Wood waste streams

Various wood waste streams can be used in engineered wood materials. Engineered wood recycling is more complicated, however, partly because of the wide use of potentially harmful formaldehyde binders – at the time of writing, MDF recycling is very unusual.

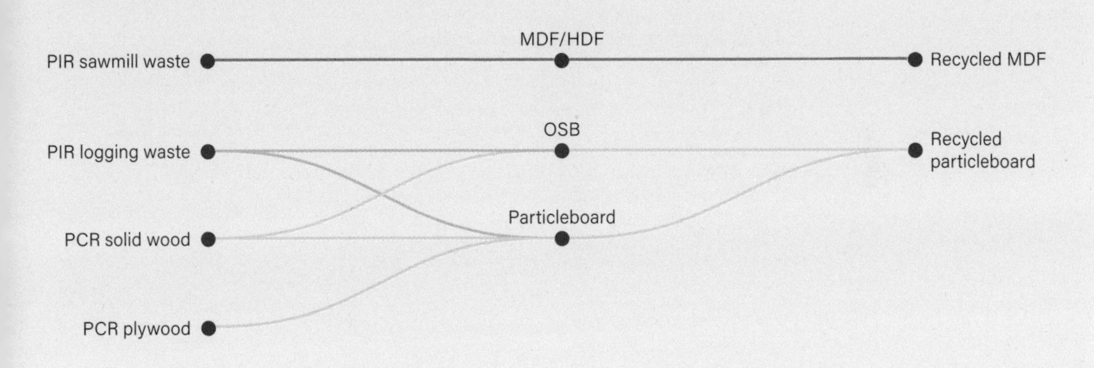

Engineered wood and formaldehyde

While the health risks of formaldehyde-based resins are well known, formaldehyde is also, rather confusingly, a natural ingredient in wood, albeit in very small quantities. No wood material, therefore, can claim to be entirely formaldehyde free, which helps to explain the criteria for certifications that restrict the use of formaldehyde and other VOCs in engineered wood.

- **No added formaldehyde (NAF)**
  No formaldehyde-based resins allowed; any formaldehyde in the engineered wood material is a natural ingredient in the wood that it was made with.

- **Ultra-low emitting formaldehyde (ULEF)**
  This certificate typically means that phenol-formaldehyde resin has been used, which has lower emissions than other types of formaldehyde-based resins. Permitted formaldehyde emission levels are 0.05 ppm for plywood, 0.09 ppm for particleboard and 0.11 ppm for MDF.

- **UL GREENGUARD and GREENGUARD Gold**
  UL GREENGUARD allows maximum formaldehyde emissions of 0.05 ppm for all engineered wood products, while UL GREENGUARD Gold is extremely strict, with only 0.0073 ppm allowed.

NAF plywood

Plywood production has traditionally used formaldehyde-based resin binders that emit toxic VOCs, but a growing number of suppliers are offering no added formaldehyde (NAF) plywood materials that use alternative, formaldehyde-free resins.

Suppliers & materials	- Panguaneta PureGlue™ NAF plywood - Dyas NAF plywood - Columbia Forest Products PureBond® NAF plywood
Raw material origin	FSC- and PEFC-certified plywoods are typically available on request. The materials listed above use petrochemical MDI resin binders, except Columbia Forest Products' PureBond® plywood, which uses a renewable plant-based adhesive.
	Panguaneta PureGlue™ NAF plywood[6]
GWP (process)	0.9 kg CO_2e / kg
GWP (biogenic)	No data
Energy use	64.5 MJ / kg
Water use	3.4 l / kg
Toxicity	The materials featured here are classified as CARB 2 NAF compliant – they do not use formaldehyde-based resin binders and have formaldehyde VOC emissions of less than 0.05 ppm.
Circularity	In theory, NAF plywood can be recycled into other engineered wood products, but many recyclers are reluctant to process plywood because it frequently contains formaldehyde. Also, the use of non-degradable resin binders – those that contain formaldehyde and others – means that plywood is not biodegradable or compostable. For an overview of wood recycling, see page 173.
Mechanical properties	Excellent strength-to-weight ratio. The strength of the material can be fine-tuned through the direction and number of veneer layers in plywood sheets.
Environmental resistance	Poor moisture and chemical resistance unless properly sealed. Plywood tends to expand and contract in response to changes in temperature and humidity, but to a lesser extent than solid wood.
Forming	Supplied in sheets that can be further formed through machining.
Finishing	The surface quality of plywood is usually graded from A to D, where A has a surface with flawless, smooth veneers and D has a very rough surface with many flaws and large knots. Plywood is compatible with common wood finishes such as oils, lacquers, paints, waxing and staining.

Röhsska Chair, designed by Fredrik Paulsen for Blå Station. The plywood used in the chair does not use a formaldehyde-based adhesive.

NAF medium-density fibreboard (MDF)

Similar to plywood production, MDF has traditionally used formaldehyde-based resin binders that emit toxic VOCs, but a growing number of suppliers are offering no added formaldehyde (NAF) plywood materials that use alternative, formaldehyde-free resins.

Suppliers & materials	- FINSA Fibracolour NAF MDF - MEDITE® CLEAR NAF MDF - Evertree MDF Green Ultimate™ NAF MDF
Raw material origin	Sawdust, bark and other post-industrial waste from sawmills. FSC- and PEFC-certified MDF available on request. The materials listed above use MDI resin binders, except Evertree MDF Green Ultimate™, which uses a renewable plant-based resin.
	FINSA Fibracolour NAF MDF[7]
GWP (process)	0.5 kg CO_2e / kg
GWP (biogenic)	-2.6 kg CO_2e / kg
Energy use	11.8 MJ / kg
Water use	82.1 l / kg
Toxicity	The MDF materials listed are classified as CARB 2 NAF compliant – they do not use formaldehyde-based resin binders and have formaldehyde VOC emissions of less than 0.11 ppm.
Circularity	Although NAF MDF is recyclable in theory, many recyclers are reluctant to process MDF because it frequently contains formaldehyde. At the time of writing, recycled MDF from the Belgian supplier UNILIN Panels is likely the only commercially available recycled MDF material. Also, the use of non-degradable resin binders – those that contain formaldehyde and others – means that MDF is not biodegradable or compostable. For an overview of wood recycling, see page 173.
Mechanical properties	While MDF has uniform strength and mechanical properties in all directions, it doesn't match the structural strength of plywood and is also considerably heavier.
Environmental resistance	Poor moisture resistance unless properly sealed. Unlike solid wood and some engineered wood materials, MDF doesn't expand and contract in response to changes in temperature and humidity.
Forming	Supplied in sheets and blocks that can be further formed through machining.
Finishing	The surface of MDF sheets is very smooth and uniform, suitable for paint and lacquers, and it can also be varnished and stained.

Montana System modular shelving. The units are made with MDF that uses a resin binder without formaldehyde.

A. JOURNAL OF WEST COAST CULINARY

NAF oriented strand board (OSB)

The development of formaldehyde-free resins for oriented strand board (OSB) production has lagged behind plywood and MDF, but a growing number of suppliers are now able to offer no added formaldehyde (NAF) OSB materials.

Suppliers & materials	- MEDITE® SMARTPLY NAF OSB - Bonzano OSB Color NAF OSB
Raw material origin	PIR timber waste such as offcuts, and crooked and knotty wood that's unsuitable for solid wood production. FSC- and PEFC-certified OSB available on request. The materials listed above use MDI resin binders.
	MEDITE® SMARTPLY NAF OSB[8]
GWP (process)	0.5 kg CO_2e / kg
GWP (biogenic)	-1.5 kg CO_2e / kg
Energy use	23.6 MJ / kg
Water use	1.8 l / kg
Toxicity	Although OSB materials are formally exempt from the CARB 2 standard, the materials listed above are classified as NAF because they do not use formaldehyde-based resin binders.
Circularity	Although NAF OSB is recyclable in theory, many recyclers are reluctant to process OSB because it frequently contains formaldehyde. Also, the use of non-degradable resin binders - those that contain formaldehyde and others - means that OSB is not biodegradable or compostable. For an overview of wood recycling, see page 173.
Mechanical properties	Strong and lightweight, with comparable performance to plywood.
Environmental resistance	Poor moisture resistance unless properly sealed. Several grades of NAF OSB materials are available with enhanced moisture and flame resistance.
Forming	Supplied in sheets and blocks that can be further formed through machining.
Finishing	OSB sheets are available in different surface qualities, from smooth to rough. However, the wood 'strands' are always clearly visible in the surface. Some suppliers offer a selection of standard colours in pre-finished OSB sheets, as well as colour matching for large orders. OSB is compatible with common wood varnishes, paints and oil treatments.

Bonzano OSB Color samples. Bonzano offers a range of colours and gloss levels in pre-finished NAF OSB sheets.

High-pressure laminates (HPLs)

Although high-pressure laminates (HPLs) are technically paper-based, they are often used in combination with engineered wood materials and typically use similar formaldehyde-based resin binders. The HPLs featured here have very low or no VOC emissions at all.

Suppliers & materials	- Abet Laminati SCS Indoor Advantage™ Gold-certified HPL - Panolam® Pionite® and Nevamar® UL GREENGUARD Gold-certified HPL - Dekodur® formaldehyde-free ECO-HPL
Raw material origin	Kraft paper with up to 20% recycled material in the core, combined with a printed virgin paper decorative outer layer. These materials are typically saturated in phenol-formaldehyde resin, although Dekodur® ECO-HPL uses an alternative, plant-based resin. FSC- and PEFC-certified HPLs are available on request. Ask suppliers to provide SCS Recycled Content Certification or other relevant certification to confirm the origin of raw materials.
	Abet Laminati PRINT HPL Thin[9]
GWP	3.7 kg CO_2e / kg
Energy use	159 MJ / kg
Water use	40 l / kg
Toxicity	Although the HPL materials from Abet Laminati and Panolam® use phenol-formaldehyde resins, both have received low VOC certification. Dekodur® ECO-HPL uses an alternative, plant-based resin. Ask other suppliers to provide SCS Indoor Advantage™, UL GREENGUARD Gold or other relevant certification for specific materials.
Circularity	HPLs are currently not recycled, and although they are paper-based, the use of non-degradable resin binders means that they are not biodegradable and cannot be composted. For an overview of wood recycling, see page 173.
Mechanical properties	Strong and stiff with excellent scratch resistance, HPLs are quite dense and heavy and are often used with a lighter backing material such as MDF for adding bulk.
Environmental resistance	Excellent moisture, UV and chemical resistance.
Forming	Supplied in sheets that can be further formed through sawing, machining and other wood forming processes.
Finishing	Available in a wide range of standard colours and finishes, with custom design and colour matching possible for large orders.

In a Box exhibition at Milano Design Week 2022. The pieces were covered with laminates from the Whimsy Collection by Arthur Arbesser x Abet Laminati.

Cork composites

Cork composite materials are typically made with post-industrial waste from the wine stopper and footwear industries, which is ground up and formed into sheets and blocks.

Suppliers & materials	- Amorim expanded corkboard - Amorim Wise UL GREENGUARD Gold-certified cork composite - Granorte cork composite
Raw material origin	Predominantly PIR cork waste from wine stopper and footwear production. The suppliers featured here offer FSC- and PEFC-certified cork. Amorim expanded corkboard is made entirely with cork, while Amorim Wise cork composite uses cork particles mixed with a petrochemical-based polyurethane binder. Ask suppliers to confirm the type of binder used in specific cork composite materials.
	Amorim expanded corkboard[10]
GWP	–1.7 kg CO_2e / kg (process and biogenic GWP combined)
Energy use	68 MJ / kg
Water use	14 l / kg
Toxicity	Pure cork as used in Amorim expanded corkboard is considered non-toxic. Amorim Wise cork composite has received UL GREENGUARD Gold certification, meaning that VOC emissions are very low. Ask cork composite suppliers to provide documentation for VOC emissions, food contact and other relevant certification based on your requirements.
Circularity	Cork composites are currently not widely recycled. Expanded corkboard is made entirely with PIR cork waste and is biodegradable, while Amorim Wise cork composite uses a polyurethane binder and cannot be composted. For an overview of wood recycling, see page 173.
Mechanical properties	Excellent compression strength, with some flexibility and good shape retention. Expanded corkboard sheets are more prone to tearing than Wise cork composite sheets.
Environmental resistance	Cork is naturally moisture resistant, but it will stain on contact with food and chemicals. Amorim Wise cork composite can be specified in moisture- and chemical-resistant grades.
Forming	Supplied in sheets and blocks that can be further formed through machining.
Finishing	Amorim expanded corkboard is available in natural colour only and has a fairly rough surface finish, while Amorim Wise cork composite is available in many different colours and finishes.

Bob Stool, designed by Michael Sodeau for Modus. The stool is made with a cork composite material from the Portuguese supplier Granorte that uses FSC-certified PIR cork waste from the wine-stopper industry.

6 Paper

Paper has a lot going for it from an environmental point of view – lightweight, renewable and widely recycled, it's also biodegradable and potentially a source of biofuel. In fact, many paper mills are partially or fully powered with waste from the papermaking process. Although timber is by far the dominant source of cellulose fibre for paper production, every plant on the planet contains cellulose and could be used for papermaking in theory, opening up a very wide range of potential sources of raw material. Paper is also a relatively straightforward material to work with – unlike metals, plastics and glass, which typically require heating and complex tooling for forming, paper can be formed using relatively simple cold-forming processes.

This is not to say that the paper industry is without environmental challenges. In many parts of the world, excessive and often illegal logging is a major problem, causing deforestation and soil erosion. The papermaking process also consumes large amounts of water, which of course adds to the water footprint of paper, but it can also be a source of pollution if wastewater is not cleaned properly before it's released back into the environment. Although mostly phased out at this stage, harsh chemicals are still used in papermaking in some parts of the world – in particular, chlorine for bleaching (see page 198).

Paper also has some intrinsic material property drawbacks that must be taken into account when using it in packaging and other product design applications. Its poor resistance to grease, water and other liquids is something of an Achilles heel; it also has poor tear resistance and overall low structural strength. This chapter features several options that address these shortcomings without compromising on recyclability or biodegradability, with a focus on packaging paper, paperboard and moulded paper pulp, as shown opposite.

There are several sustainability standards and certifications for paper. On the raw material side, both the Forest Stewardship Council (FSC) and the Programme for the Endorsement of Forest Certification (PEFC) offer certifications for paper made with responsibly sourced wood as well as recycled paper content, while the EU Ecolabel for paper sets stringent requirements for the origin of raw materials, use of hazardous chemicals, energy efficiency and waste management in paper production. In addition, ECOLOGO© certification, from US-based UL Solutions, includes specific standards for printing and paper coatings.

Paper materials

All plants contain cellulose fibre, the basic building block of paper, so in principle any plant can be used to make paper. From sheet materials to pulp mouldings, this book features a wide range of paper materials for packaging and other product applications.

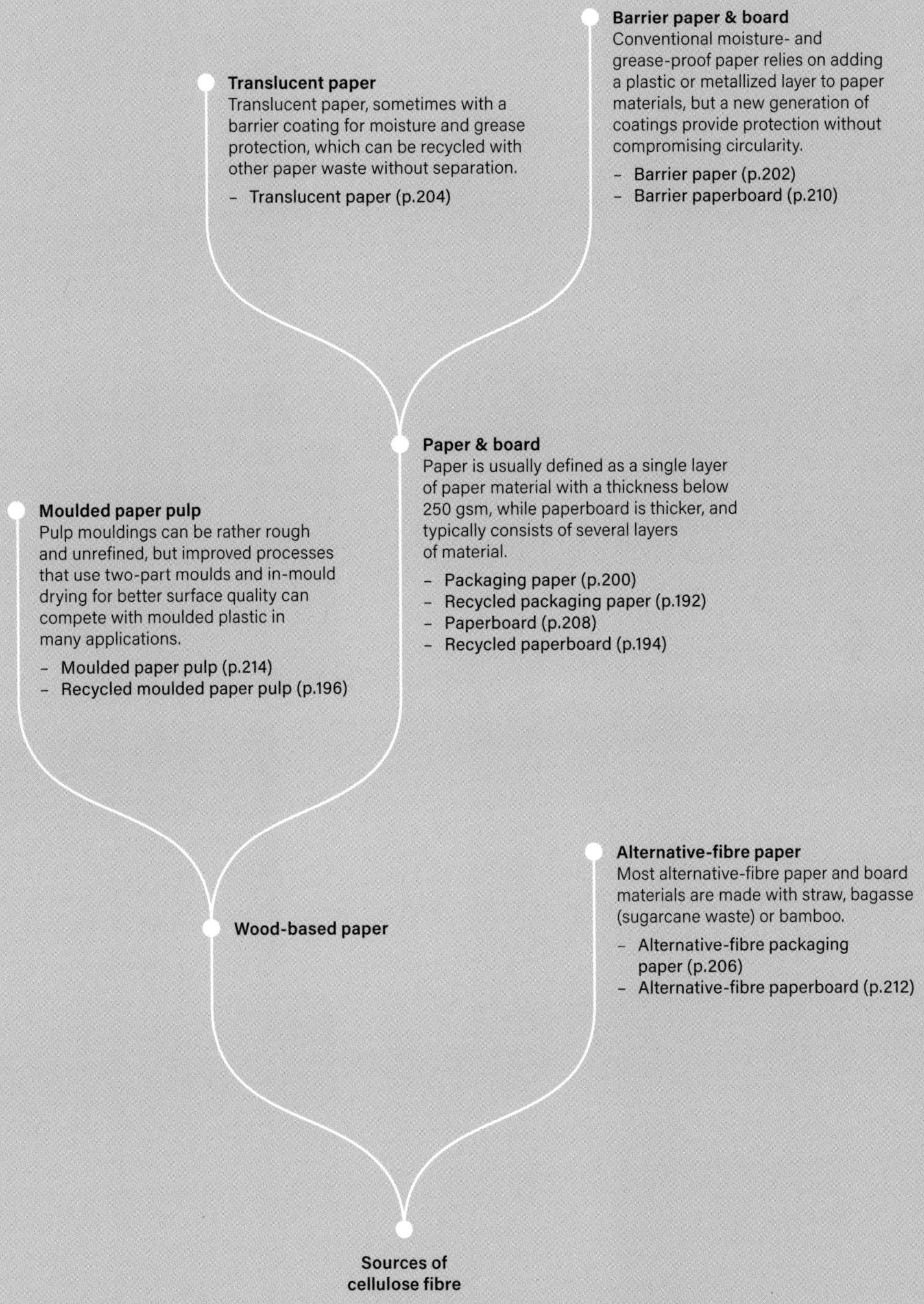

Barrier paper & board
Conventional moisture- and grease-proof paper relies on adding a plastic or metallized layer to paper materials, but a new generation of coatings provide protection without compromising circularity.
- Barrier paper (p.202)
- Barrier paperboard (p.210)

Translucent paper
Translucent paper, sometimes with a barrier coating for moisture and grease protection, which can be recycled with other paper waste without separation.
- Translucent paper (p.204)

Paper & board
Paper is usually defined as a single layer of paper material with a thickness below 250 gsm, while paperboard is thicker, and typically consists of several layers of material.
- Packaging paper (p.200)
- Recycled packaging paper (p.192)
- Paperboard (p.208)
- Recycled paperboard (p.194)

Moulded paper pulp
Pulp mouldings can be rather rough and unrefined, but improved processes that use two-part moulds and in-mould drying for better surface quality can compete with moulded plastic in many applications.
- Moulded paper pulp (p.214)
- Recycled moulded paper pulp (p.196)

Alternative-fibre paper
Most alternative-fibre paper and board materials are made with straw, bagasse (sugarcane waste) or bamboo.
- Alternative-fibre packaging paper (p.206)
- Alternative-fibre paperboard (p.212)

Wood-based paper

Sources of cellulose fibre

'I want to give paper dignity and value.'

An interview with Riccardo Cavaciocchi, architect and founder of Paper Factor – a striking solid surface material made with cellulose fibre from a wide range of raw material sources in a former wine factory in Lecce, southern Italy. Find out more about Riccardo and Paper Factor at paperfactor.com

When I first came across Paper Factor, it was the beauty of the material that drew me in. Then when I started to understand more about the process and how the material is made I was even more intrigued. Could you tell me a bit about the background of the material, as well as your own?

My mother, Lidiana Miotto, is a conservationist with a long career of restoring works of art in public and private collections in Italy (Centro Restauro Materiale Cartaceo). Around the time that I was born she invented the compound that is the basis of Paper Factor, so you could say that I've been around this material my whole life. I love the hands-on quality of paper, and Paper Factor is very much based on experimentation and trying different things to see what happens.

Compared with other materials that require complex equipment, most of the equipment that we use at Paper Factor is more or less what you would find in a domestic kitchen. We also place a strong emphasis on contemporary technology. The main ingredients are a combination of FSC-certified virgin paper (fibres) and recycled paper, but beyond that we experiment with a wide range of waste-based raw materials, including waste and by-products from growing vegetables, as well as wine, chocolate, tobacco and olive oil production, but also fashion and textile waste, such as post-consumer recycled jeans.

Movable Mini Tables Set,
by Paper Factor.

Paper is an interesting material from a sustainability perspective, being renewable and recyclable. What is your personal take on this in relation to Paper Factor?

For me, Paper Factor has always been about preservation. The way I see it, the Paper Factor process is essentially a way of reducing waste and turning it into something useful. This thinking is in my DNA, not just from my mother's work as a conservationist, but also as a way of life. If something is broken, you fix it rather than throwing it away. I run Paper Factor based on these principles – for example, we are always keeping a very close eye on drying times for our materials so that we can conserve energy and be more efficient.

As a designer, my first thought about Paper Factor was that it is so different from my preconceptions about paper. My next thought was, 'What can I design with this material?'

As far as I'm concerned, there's no real limit to how Paper Factor could be used – so far, it's been used in furniture, lighting and other interior accessories.

In terms of aesthetics, there's an inherent element of imperfection in paper that's really appealing to me – if you look at paper closely enough you can see the randomness of the fibre in the surface. In a way, what we do at Paper Factor is like alchemy, extracting the inner beauty of paper and turning it into a material that can be used in applications that are more durable than you would normally expect from it. You could say that I want to give paper dignity and value as an architectural material.

Paper Factor samples.

Paper recycling

It's estimated that just over 50 per cent of the total amount of paper produced in 2019 was made with recycled fibre,[1] very much reducing the amount of virgin raw materials, energy, water and chemicals needed in the global paper industry.

While it's extremely positive that there's such a large demand for recycled paper, paper recycling does have a few disadvantages that designers should be aware of in order to get the most out of the material.

Recycled paper is essentially re-pulped, meaning that the cellulose fibre is shredded and shortened as a result. This makes recycled paper weaker than virgin paper, which can mean that a thicker gauge of recycled paper material is needed to match the performance of virgin paper. Also, because the cellulose fibre is shortened every time it's recycled, paper can only be recycled four to six times before the fibre becomes too short. According to the Confederation of European Paper Industries (CEPI), paper is recycled on average 2.4 times in Europe. This means that there's an intricate relationship between the production of virgin and recycled paper, as the pool of recycled material needs to be continually renewed with virgin paper in order not to run out.

Most paper recycling mills have certain limitations around the type of paper materials that they can process. Printing inks, water-soluble coatings and adhesives can be separated in most paper recycling mills, but more complex paper materials, such as the kind of paper-based multilayered laminates that are used for milk cartons, can only be processed by specialist paper recyclers. The paper materials, coatings and adhesives featured in this book all belong to the first category – namely those materials that can be handled by standard paper recycling mills.

The diagram shown at the bottom of the page opposite gives an overview of the paper recycling process. First, paper waste is collected and sorted into separate categories, including packaging paper and paperboard, cardboard, newspapers and magazines. Recycled paper for food contact needs special treatment during the recycling process. Paper waste that's wet or contains excessive food residues cannot be recycled, but it's still suitable for composting. Recycled paper used in graphic paper and other high-quality applications is further cleaned and de-inked before it's sent for further processing into paper at a paper mill.

Both the FSC and PEFC offer certifications for recycled paper. FSC Recycled is for 100 per cent recycled paper, and FSC Mix is for paper that contains a mix of virgin and recycled fibre. The PEFC Recycled label applies to paper made with at least 70 per cent recycled fibre. Additional credentials include SCS Global Services Recycled Content Certification, which specifies the type and percentage of recycled material.

Global paper recycling rates

According to the Bureau of International Recycling (BIR), about 412 million tonnes of paper and board material were produced in 2019, with just over half made from recycled fibre. Asian and Latin American paper mills incorporated the highest amount of recycled fibre in paper and board production at around 70%, followed by Europe at around 55% and the rest of the world at about 40% on average.[2]

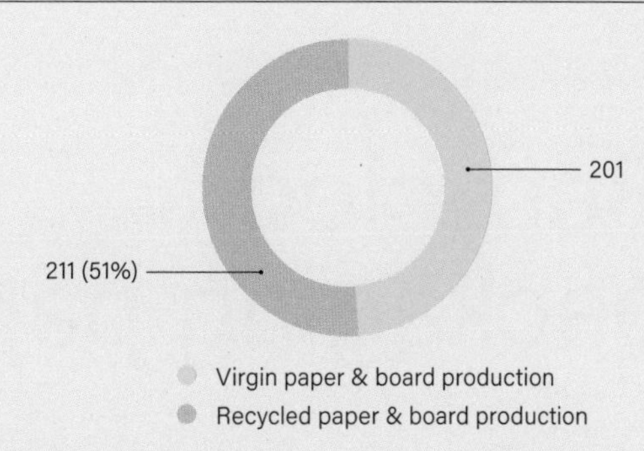

Uses for recycled paper

Recycled paper, measured in million tonnes. The largest volumes of recycled paper are used in packaging, with a global average recycled content rate of about 70%, followed by newsprint, at around 65%. At the other end of the scale, graphic and printing paper uses only around 9% recycled material.[3]

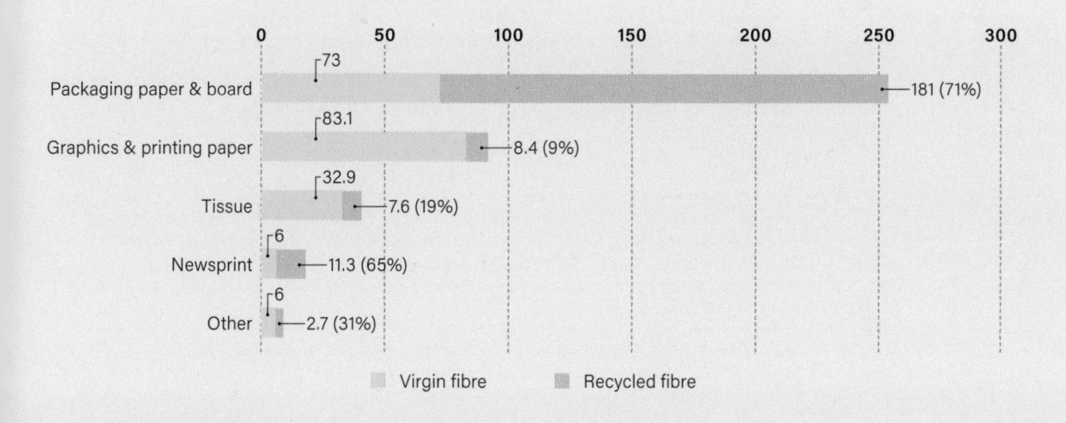

Paper waste streams

High-quality graphic and printing paper is usually de-inked and bleached separately, which isn't necessary for other categories of paper waste. Soiled and greasy paper will contaminate other waste, so this should be discarded and preferably composted. The same goes for paper that has been recycled more than four to six times, to the point where the cellulose fibre has become too short.

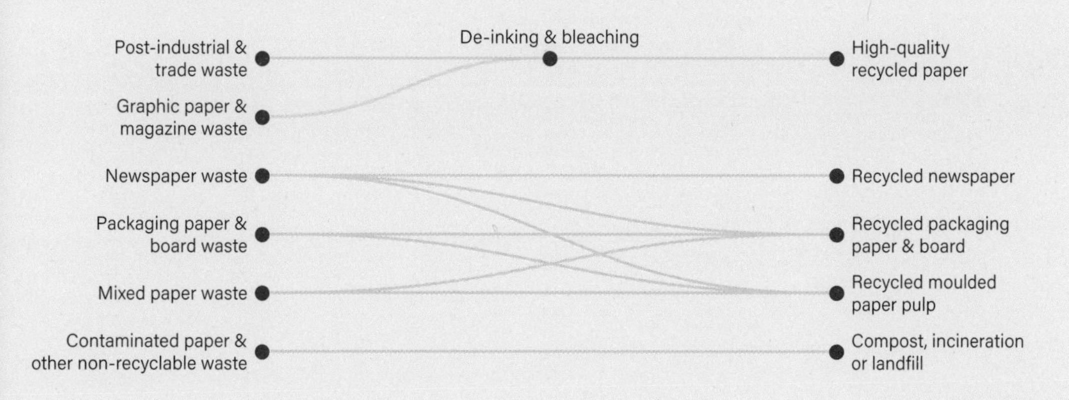

Recycled packaging paper

Packaging paper is widely recycled, with a significantly lower environmental impact compared with virgin material.

Suppliers & materials	- ArjoWiggins Cyclus Pack and Cocoon Pack recycled packaging paper - LEIPA Classic White C, Lux PLUS and Spirit White recycled packaging paper - Mondi EcoVantage recycled packaging paper	
Raw material origin	A combination of PIR and PCR paper waste mixed with virgin fibre. Ask suppliers to confirm the exact ratio of recycled raw materials, as well as any relevant certification, such as FSC Recycled and PEFC Recycled.	
	ArjoWiggins Cyclus Pack[4]	**Virgin FSC packaging paper**[5]
GWP	0.5 kg CO_2e / kg	1.3 kg CO_2e / kg
Energy use	No data	20.4 MJ / kg
Water use	No data	480 l / kg
Toxicity	Recycled paper can contain traces of petrochemical-based inks from certain paper waste, such as newsprint. Petrochemical-based ink residues in recycled paper packaging materials have been shown to contaminate food, so ask paper suppliers to confirm that specific recycled papers are safe for food contact. Also confirm that bleached recycled paper uses an elemental chlorine-free process.	
Circularity	Recyclable with other paper waste, so long as the material is kept dry and free from excessive food waste. Soiled and other non-recyclable paper waste can be composted at home or in industrial composting facilities instead. Printing inks, water-based adhesives and recyclable paper-based labels do not affect the recyclability or biodegradability of the paper. For an overview of recycling rates and circular design guidelines for paper, see page 191.	
Mechanical properties	The cellulose fibre in paper is weakened during the recycling process, so recycled papers typically have lower **burst strength** and bulk compared to virgin material. Ask suppliers to provide a technical data sheet for specific recycled packaging papers to benchmark performance with virgin paper alternatives.	
Environmental resistance	Recycled packaging papers without a barrier coating have poor resistance to moisture and grease. Due to its flammability, paper should not be used in high-temperature applications.	
Forming	Compatible with common paper forming and assembly processes, such as various cutting processes and folding. Joining with water-based adhesives doesn't affect the recyclability and biodegradability of the material.	
Finishing	Recycled packaging papers are available in both bleached (white) and brown, sometimes grey, unbleached versions. Some recycled papers have black spots from ink residue in the surface, which can be a decorative feature. Most are straightforward to print, although certain materials may be optimized for specific printing processes and inks. Low-migration, petrochemical-free inks should be used to improve the recyclability and biodegradability of the paper.	

By Humankind's shipping bag, made from recycled packaging paper.

6.0 × 9.0 in.
Recycled
Paper Parcel

by
Humankind

Recycled paperboard

While recycled paperboard has a lower environmental impact than virgin material, it's also typically weaker, which may require design adjustments for applications that require structural strength.

Suppliers & materials	- ArjoWiggins Teknocard recycled paperboard - LEIPA Board recycled paperboard - Westrock ReNew100® recycled paperboard	
Raw material origin	Combination of PIR and PCR paper waste mixed with virgin fibre. Ask suppliers to confirm the exact ratio of recycled raw materials, as well as any relevant certification, such as FSC Recycled and PEFC Recycled.	
	ArjoWiggins Teknocard[6]	**FSC paperboard**[7]
GWP	0.7 kg CO_2e / kg	1.2 kg CO_2e / kg
Energy use	No data	18.5 MJ / kg
Water use	No data	2,490 l / kg
Toxicity	Recycled paperboard can contain traces of petrochemical-based inks from certain paper waste, such as newsprint. Petrochemical-based ink residues in recycled paper packaging materials have been shown to contaminate food, so ask paper suppliers to confirm that specific recycled papers are safe for food contact. Also confirm that bleached recycled paper uses an elemental chlorine-free process.	
Circularity	Recyclable with other paper waste, so long as the material is kept dry and free from excessive food waste. Soiled and other non-recycled paper waste can be composted at home or in industrial composting facilities. Printing inks, water-based adhesives and recyclable paper-based labels do not affect the recyclability or biodegradability of the paper. For an overview of recycling rates and circular design guidelines for paper, see page 191.	
Mechanical properties	The cellulose fibre in paper is weakened during the recycling process, so recycled paperboard materials typically have lower tensile strength, rigidity and bulk compared to virgin materials. Ask suppliers to provide a technical data sheet for specific recycled packaging papers to benchmark performance with virgin paperboard alternatives.	
Environmental resistance	Recycled paperboard materials without a barrier coating have poor resistance to moisture and grease. Due to its flammability, paper should not be used in high-temperature applications.	
Forming	Compatible with common paper forming and assembly processes, such as various cutting processes and folding. Joining with water-based adhesives doesn't affect the recyclability and biodegradability of the material.	
Finishing	Recycled paperboard materials are available in both bleached (white) and brown, sometimes grey, unbleached versions. Some recycled paperboard has black spots from ink residue in the surface, which can be a decorative feature. Most recycled papers are straightforward to print, although certain materials may be optimized for specific printing processes and inks. Low-migration, petrochemical-free inks should be used to improve the recyclability and biodegradability of the paper.	

Sumpan earphones by Urban Ears.
The packaging is made with recycled paperboard.

Recycled moulded paper pulp

Most paper pulp moulded products, such as egg cartons, are made with recycled raw materials, although some applications that require smoother surfaces and better colourability may use virgin material (see page 214).

Suppliers & materials	- Pulp-Tec recycled moulded paper pulp - UFP Technologies recycled moulded paper pulp - TRIDAS® recycled moulded paper pulp
Raw material origin	Typically newspaper waste. Ask suppliers to confirm the exact ratio of recycled raw materials, as well as any relevant certification, such as FSC Recycled and PEFC Recycled.
	Recycled moulded paper pulp[8]
GWP	0.4 kg CO_2e / kg
Energy use	0.5 MJ / kg
Water use	0.5 l / kg
Toxicity	Recycled moulded paper pulp almost certainly contains traces of petrochemical-based inks, as these inks are commonly used in newsprint, the main source of raw material for recycled moulded paper pulp. Petrochemical-based ink residues in paper packaging materials have been shown to contaminate food, so ask moulded paper pulp suppliers to confirm food-contact approval.
Circularity	Recyclable with other paper waste, so long as the material is kept dry and free from excessive food waste. Soiled and otherwise non-recyclable moulded paper pulp can be composted instead. Printing inks, water-based adhesives and recyclable paper-based labels do not affect the recyclability or biodegradability of moulded paper pulp.
Mechanical properties	While recycled moulded paper pulp offers good energy absorption, it has lower tensile and compression strength and dimensional stability compared to virgin material due to its shorter, recycled fibres. Its part strength and rigidity can be improved by adding ribs, selectively thicker wall sections and other features.
Environmental resistance	Poor resistance to moisture and grease. Moulded paper pulp should not be used in high-temperature applications due to its flammability.
Forming	Recycled moulded pulp is typically formed using either simple compression moulding or thermoforming, with thermoforming allowing for more complex shapes and better surface quality.
Finishing	Available in both bleached and brown, sometimes grey, unbleached versions. The surface of recycled moulded paper pulp parts can be very rough, making it difficult to print directly onto the material. The pulp can be dyed for solid colour applications, although any impurities in the material, such as ink residues, will be visible.

Green Clean toothbrush packaging by Jordan. The moulded pulp packaging is made with recycled paper.

SOFT
click
CHANGE
2 HEADS +
1 HANDLE
Jordan*
SCANDINAVIAN
QUALITY

Virgin paper

Although timber is the main raw material for paper production, cellulose fibre - paper's basic building block - exists in every cell of every plant, making it the most abundant organic material on earth. A quick search for paper made with alternatives to timber will reveal examples made with grass, cherry stones, algae, wheat husks, waste from beer breweries and more.

Cellulose content varies from plant to plant – in cotton it's about 90 per cent, while wood contains about 45 per cent. A plant with a high cellulose content is not necessarily the obvious choice for papermaking, however. The reason that wood is so dominant is the relative ease of growing and harvesting huge numbers of trees. Nevertheless, there is intense development of alternative sources for paper production, with agricultural waste being one of the most promising areas. Stubble burning (burning what remains after grains are harvested) is a major cause of air pollution in India, so straw-based paper is a way to divert agricultural waste that will otherwise likely be burnt.

There are two dominant industrial papermaking processes. The first is mechanical pulping, which grinds timber and other raw materials down mechanically to make pulp. As a result, substances other than cellulose that are in the raw material will also end up in the paper. In the case of timber, this is mainly another organic compound called lignin. Lignin has several negative impacts on the quality of paper, including reduced strength (it weakens the bond between cellulose fibres), as well as causing paper to yellow gradually when exposed to oxygen and sunlight.

The second process is chemical pulping, aka kraft pulping. Instead of grinding the raw material, it uses hot water and chemicals to extract cellulose fibre and separate other compounds like lignin. Kraft paper consists almost entirely of cellulose fibre, making it stronger than paper made with mechanical pulping, as well as resistant to yellowing.

About 80 per cent of global production uses the kraft process, but mechanical pulping is still useful. The latter has a yield rate of about 95 per cent and it's often used to make the core material in paperboard, sandwiched with kraft paper.

Paper made with both processes is often bleached – otherwise the end result will be brown or beige, with visible fibre in the surface. In the past, almost all paper was bleached using chlorine that was often released with wastewater from paper mills back into nearby rivers and lakes. Although chlorine is still used in some parts of the world, wastewater treatment is subject to stronger restrictions and other bleaching processes are now more common, including elemental chlorine free (ECT) bleaching and totally chlorine free (TCF) bleaching. Ask suppliers to confirm which process they use, as well as any other relevant certification – such as FSC and PEFC for responsible sourcing of timber, and the EU Ecolabel, which sets strict standards for the use of chemicals, energy efficiency and waste management in the paper industry.

Global virgin paper production

Some 189 million tonnes, or 94%, of virgin paper production in 2020 used wood-based cellulose fibre. Unlike many other sources of cellulose fibre, such as agricultural crops and, by extension, agricultural waste, wood is a reliable, year-round resource. In addition, the forest industry has a well-established infrastructure for harvesting, transporting and storing wood raw materials, which doesn't exist to the same extent (or at least not yet) for alternative raw materials.[1]

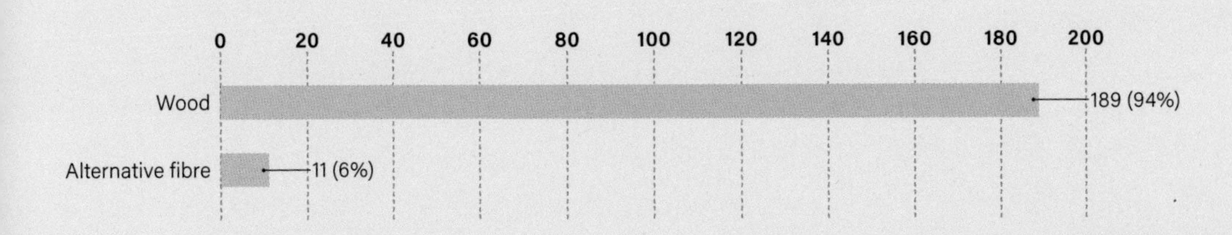

Alternative-fibre paper production

Although volumes of alternative-fibre paper are currently relatively low at 11 million tonnes annually, it can be a valuable approach to deal with over-reliance on the world's forests for paper and timber production, as well as other environmental issues such as stubble burning – a major cause of pollution in areas where the practice is widespread.[2]

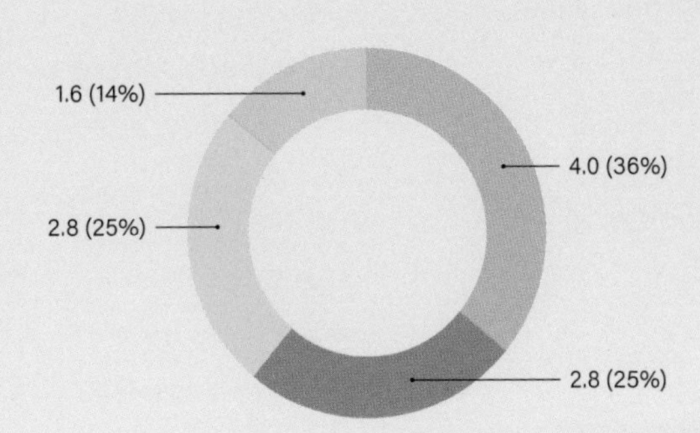

Paper certifications

Several certifications exist for different paper raw materials, with the following being the most common.

- **Wood-based paper**
  As for other forest products, PEFC and FSC both offer paper certification. Both certificates are available for paper made with wood from responsibly managed forests, as well as recycled raw materials. About 11% of global forests are PEFC, FSC or PEFC/FSC double-certified, a ratio that should be roughly reflected in global wood-based paper production.

- **Bamboo paper**
  PEFC and FSC also offer certification for bamboo forests and materials. At the time of writing, it has not been possible to establish statistics for the percentage of global bamboo production that's certified.

- **Bagasse paper**
  Bagasse is made with waste sugarcane stalks, which are pulped and made into paper. Bonsucro is the leading international certification for sugarcane production, so you should confirm that bagasse paper is made with waste from Bonsucro-certified plantations. According to Bonsucro, globally about 5% of sugarcane plantations are certified, which should be reflected in the ratio of certified bagasse paper.

Packaging paper

Several suppliers offer packaging paper that, in combination with heat-sealing adhesives, can be an alternative to plastics in flexible packaging applications that do not require moisture and grease resistance.

Suppliers & materials	- Arctic Paper Munken Kraft packaging paper - UPM UniquePack™ and Solide™ packaging paper - Billerud Axello packaging paper - Henkel LOCTITE LIOFOL recyclable and biodegradable heat-sealing adhesive
Raw material origin	Responsibly sourced timber from managed forests. Ask suppliers to confirm relevant certification, including FSC, PEFC and EU Ecolabel. Several paper suppliers can trace the origin of raw materials used on request. Heat-sealing adhesives for paper are likely to contain petrochemical-based ingredients, so ask suppliers to confirm the origin and type of ingredients used.
	Arctic Paper Munken Kraft packaging paper[3]
GWP	0.5 kg CO_2e / kg
Energy use	9.4 MJ / kg
Water use	No data
Toxicity	Paper is considered non-toxic, but the suitability of specific paper materials for food-contact applications should be confirmed, as they may react in non-toxic but unexpected ways, such as altering the colour or taste of foods. None of the packaging papers in this book use elemental chlorine for bleaching. Heat-sealing adhesives for food-contact applications are available – ask suppliers to recommend suitable grades for specific applications.
Circularity	Recyclable with other paper waste, so long as it's kept dry and free from excessive food waste. Contaminated packaging paper that cannot be recycled can be composted at home or in industrial composting facilities instead. Printing inks, recyclable paper-based labels and the type of heat-sealing adhesives featured here do not affect the recyclability or biodegradability of packaging paper. For an overview of recycling rates and circular design guidelines for paper, see page 191.
Mechanical properties	There is a direct connection between the thickness and strength of packaging papers, measured in burst strength. Ask suppliers to help identify a suitable paper weight for specific applications.
Environmental resistance	Poor resistance to moisture and grease. Paper that's made using the chemical pulping process – also known as kraft paper – has better UV resistance than paper that's made using mechanical pulping. Due to its flammability, paper should not be used in high-temperature applications.
Forming	Compatible with common paper forming and assembly processes, such as various cutting processes and folding. Packaging paper with a compatible heat-sealing adhesive can often be formed and sealed on the manufacturing lines used for flexible plastic film packaging.
Finishing	Available in both bleached (white) and brown, unbleached versions. Both alternatives can be printed, but certain materials may be optimized for specific printing processes and inks. Low-migration, petrochemical-free inks should be used to improve the recyclability and biodegradability of the paper.

Ritter Sport Mini packaging. This paper pouch is coated with a biodegradable, recyclable and heat-sealable adhesive, meaning that the paper can be formed and sealed on the same type of manufacturing lines that are used to make plastic packaging pouches.

PAPER BAG
CHOCOLATE
9 PIECES
Ritter SPORT
mini
MIX
Ritter SPORT
CORNFLAKES
Ritter SPORT
PRALINE
Ritter SPORT
ALPINE MILK CHOCOLATE
Ritter SPORT
BUTTER BISCUIT
Ritter SPORT
HAZELNUTS

Barrier paper

The latest generation of barrier papers can replace plastics in many flexible packaging applications, offering excellent moisture and grease resistance without compromising the recyclability and biodegradability of the material.

Suppliers & materials	- UPM Asendo™ barrier paper - Sappi Guard Nature barrier paper - Mondi AegisPaper barrier paper
Raw material origin	Responsibly sourced timber from managed forests. Ask suppliers to confirm relevant certification, including FSC, PEFC and EU Ecolabel. Several paper suppliers can trace the origin of raw materials used on request. Confirm the exact type of barrier coating with suppliers – ideally it should be made with renewable and sustainably sourced raw materials.
	UPM Confidio™ barrier paper[4]
GWP	0.4 kg CO_2e / kg
Energy use	1.9 MJ / kg
Water use	No data
Toxicity	The barrier papers featured here are free from fluoropolymers, PVDC and other potentially harmful materials. Ask suppliers to confirm relevant food-contact certification such as FDA and BfR. None of the barrier papers in this book use elemental chlorine for bleaching.
Circularity	Recyclable in the normal paper waste stream, but specific barrier coatings may be unsuitable for composting. Ask suppliers to confirm relevant biodegradability certification, such as OK Compost HOME and INDUSTRIAL. For an overview of recycling rates and circular design guidelines for paper, see page 191.
Mechanical properties	There is a direct connection between the thickness and strength of packaging papers, which is measured in burst strength. Ask suppliers to help identify a suitable paper weight for specific applications.
Environmental resistance	Good resistance to moisture and grease, and suppliers will often guarantee performance for a specific time frame. Paper that's made using chemical pulping – also known as kraft paper – has better UV resistance than paper made using mechanical pulping. Due to its flammability, paper should not be used in high-temperature applications.
Forming	Compatible with common paper forming and assembly processes, such as various cutting processes and folding. Joining with water-based adhesives doesn't affect recyclability and biodegradability. Some barrier papers are heat-sealable and can be formed and sealed on the manufacturing lines used for flexible plastic film packaging.
Finishing	Barrier papers are available in both bleached (white) and brown, unbleached versions. Both alternatives can be printed, but certain materials may be optimized for specific printing processes and inks. Low-migration, petrochemical-free inks should be used to improve the recyclability and biodegradability of the paper.

Tony's Chocolonely chocolate wrapper. The wrapper uses a barrier paper with a mineral-based coating that doesn't impact the biodegradability or recyclability of the paper.

TONY'S
CHOCOLONELY
almond honey nougat
together we'll make chocolate 100% slave free
Tony's Chocolonely

Translucent paper

Translucent paper offers a degree of transparency, which can make it an alternative to clear plastic in many applications, especially when combined with a barrier coating for moisture and grease resistance.

Suppliers & materials	- UPM Solide™ Lucent translucent paper - Ahlstrom-Munksjö Cristal™ translucent paper - Seaman Paper Vela translucent paper
Raw material origin	Responsibly sourced timber from managed forests. Ask suppliers to confirm relevant certification, including FSC, PEFC and EU Ecolabel. Several paper suppliers can trace the origin of raw materials used on request. If the material uses a barrier coating, confirm the type and origin of raw materials with suppliers.
	UPM Solide™ Lucent translucent paper[5]
GWP	0.46 kg CO_2e / kg
Energy use	10 MJ / kg
Water use	No data
Toxicity	None of the translucent papers in this book use elemental chlorine for bleaching. If a barrier coating is used, confirm that it's free of fluoropolymers, PVDC and other potentially harmful ingredients. Also ask suppliers to confirm relevant food-contact certifications, such as FDA and BfR.
Circularity	Recyclable with other paper waste, but if a barrier coating is used it may be unsuitable for composting. Ask suppliers to confirm relevant biodegradability certification, such as OK Compost HOME and INDUSTRIAL. For an overview of recycling rates and circular design guidelines for paper, see page 191.
Mechanical properties	There is a direct connection between the thickness and strength of packaging papers, which is measured in burst strength. Ask suppliers to help identify a suitable paper weight for specific applications.
Environmental resistance	Paper with a barrier coating has good resistance to moisture and grease, and suppliers will often guarantee performance within a specific time frame. Due to its flammability, translucent paper should not be used in high-temperature applications.
Forming	Compatible with common paper forming and assembly processes, such as various cutting processes and folding. Joining with water-based adhesives doesn't affect recyclability and biodegradability. Some translucent papers are heat-sealable and can often be formed and sealed on the manufacturing lines used for flexible plastic film packaging.
Finishing	Thickness has a direct impact on translucency – the thicker the material, the more opaque it will be. Translucent barrier papers can be printed, but certain materials may be optimized for specific printing processes and inks. Low-migration, petrochemical-free inks should be used to improve the recyclability and biodegradability of the paper.

Vela translucent paper bag by Seaman Paper. This uses uncoated paper suitable for soft goods, documents and similar packaging applications.

This bag is
made of paper.
Switching from plastic to paper is a great way for a brand to demonstrate their commitment to sustainable packaging. Paper is made from renewable resources and is one of the most highly recycled materials in the world.
please recycle this bag.

Vela

FSC

Recyclable

Made of Paper

www.vela.eco

Alternative-fibre packaging paper

Using alternative sources of cellulose fibre for paper production has the potential to reduce agricultural waste and emissions from stubble burning, a major source of pollution around the world.

Suppliers & materials	- PaperWise straw-based packaging paper - PropalPack bagasse packaging paper	
Raw material origin	PaperWise works with farmers and local paper mills in India to recover agricultural waste such as stubble, which would otherwise likely be burnt, causing severe local pollution. PropalPack uses bagasse, the leftover waste after sugarcane has been crushed, from farmers in Colombia to produce their packaging papers.	
	PaperWise straw-based paper[6]	**Virgin FSC packaging paper**[7]
GWP	0.7 kg CO_2e / kg	1.3 kg CO_2e / kg
Energy use	No data	20.4 MJ / kg
Water use	No data	480 l / kg
Toxicity	Alternative fibre paper is considered non-toxic, but the suitability of specific paper materials for food-contact applications should be confirmed with suppliers, as it can react in non-toxic but unexpected ways, such as altering the colour or taste of foods. None of the alternative-fibre packaging papers in this book use elemental chlorine for bleaching.	
Circularity	Recyclable with other paper waste, so long as the material is kept dry and free from excessive food waste. Contaminated paper that cannot be recycled can be composted at home or in industrial composting facilities instead. Printing inks, water-based adhesives and recyclable paper-based labels do not affect the recyclability or biodegradability of the paper. For an overview of recycling rates and circular design guidelines for paper, see page 191.	
Mechanical properties	There is a direct connection between the thickness and strength of packaging papers, which is measured in burst strength. Ask suppliers to help identify a suitable paper weight for specific applications.	
Environmental resistance	Poor resistance to moisture and grease. Paper that's made using chemical pulping - also known as kraft paper - has better UV resistance than paper made using mechanical pulping. Due to its flammability, paper should not be used in high-temperature applications.	
Forming	Compatible with common paper forming and assembly processes, such as various cutting processes and folding. Joining with water-based adhesives doesn't affect recyclability or biodegradability.	
Finishing	Available in bleached (white) and brown, unbleached versions. Some alternative-fibre papers have visible fibre and particles in the surface. The materials are printable, but certain materials may be optimized for specific printing processes and inks. Low-migration, petrochemical-free inks should be used to improve the recyclability and biodegradability of the paper.	

Soap bar by Loeze. The packaging paper is made with agricultural waste straw sourced in India by the Dutch paper supplier PaperWise.

loeze®
ORGANIC
HANDMADE SOAP

Big Boost
Oppeppende sheaboter
en de stimulerende geur
van citroengras

Paperboard

Paperboard is typically made with several layers of material for added stiffness, making it suitable for applications that require improved strength, such as structural packaging.

Suppliers & materials	- MetsäBoard: MetsäBoard Natural FBB, MetsäBoard Classic FBB, MetsäBoard Pro FBB OBAfree, MetsäBoard Pro FBB Bright and MetsäBoard Prime FBB Bright paperboards - Billerud CrownBoard paperboard - Holmen Iggesund Incada® paperboard - Sappi Algro and Atelier paperboard
Raw material origin	Responsibly sourced timber from managed forests. Ask suppliers to confirm relevant certification, including FSC, PEFC and EU Ecolabel. Several paper suppliers can trace the origin of raw materials used on request.
	Billerud CrownBoard paperboard[8]
GWP	0.5 kg CO_2e / kg
Energy use	22 MJ / kg
Water use	45 l / kg
Toxicity	Paper is considered non-toxic, but the suitability of specific paper materials for food-contact applications should be confirmed with suppliers, as it can react in non-toxic but unexpected ways, such as altering the colour or taste of foods. None of the paperboard materials in this book use elemental chlorine for bleaching.
Circularity	Recyclable with other paper waste, so long as the material is kept dry and free from excessive food waste. Contaminated paperboard that cannot be recycled can be composted at home or in industrial composting facilities instead. Printing inks, water-based adhesives and recyclable paper-based labels do not affect the recyclability or biodegradability of paperboard. For an overview of recycling rates and circular design guidelines for paper, see page 191.
Mechanical properties	Tough and rigid, paperboard materials also offer good compression strength. Request a technical data sheet for specific materials, weights and grades.
Environmental resistance	Paperboard materials without a barrier coating have poor resistance to moisture and grease. Paper that is made using chemical pulping – also known as kraft paper – has better UV resistance than paper made using mechanical pulping. Due to its flammability, paper should not be used in high-temperature applications.
Forming	Compatible with common paper forming and assembly processes, such as various cutting processes and folding. Joining with water-based adhesives doesn't affect the recyclability and biodegradability of paperboard.
Finishing	Paperboard materials are available in both bleached (white) and brown, unbleached versions. Both alternatives can be printed, but certain materials may be optimized for specific printing processes and inks. Low-migration, petrochemical-free inks should be used to improve the recyclability and biodegradability of the paperboard.

Arctic Blue Gin packaging. This box is finished with Envirofoil®, a decorative foil from the US-based supplier Hazen that contains no plastic and only a small amount of aluminium, so doesn't affect the recyclability and biodegradability of the paperboard.

ARCTIC BLUE
GIN
500ML 46.2% VOL.
HANDCRAFTED IN FINLAND
FROM ARCTIC BOTANICALS
AND PURE SPRING WATER

Barrier paperboard

Barrier paperboard has good strength and stiffness, combined with excellent moisture and grease resistance.

Suppliers & materials	- MetsäBoard Prime FBB EB barrier paperboard - Stora Enso Cupforma Natura™ Bio barrier paperboard - Holmen Iggesund Invercote Bio E barrier paperboard
Raw material origin	Responsibly sourced timber from managed forests. Ask suppliers to confirm relevant certification, including FSC, PEFC and EU Ecolabel. Several paper suppliers can trace the origin of raw materials used on request. The barrier coatings used in the materials featured here are based on renewable raw materials; ask suppliers to confirm the exact type of coating and any relevant certification.
	MetsäBoard Prime FBB EB barrier paperboard[9]
GWP	0.16 kg CO_2e / kg
Energy use	4.4 MJ / kg
Water use	No data
Toxicity	The barrier paperboard materials featured here are free from fluoropolymers, PVDC and other potentially harmful materials. Ask suppliers to confirm relevant food-contact certificates such as FDA and BfR. None of the barrier paperboard materials in this book use elemental chlorine for bleaching.
Circularity	Recyclable with other paper waste, but specific barrier coatings may be unsuitable for composting. Ask suppliers to confirm relevant biodegradability certification, such as OK Compost HOME and INDUSTRIAL. For an overview of recycling rates and circular design guidelines for paper, see page 191.
Mechanical properties	Tough and rigid, paperboard materials also offer good compression strength. Request a technical data sheet for specific materials, weights and grades.
Environmental resistance	Barrier paperboard materials have good resistance to moisture and grease. Suppliers will often guarantee performance for a specific time frame. Paper that's made using chemical pulping – also known as kraft paper – has better UV resistance than paper made using mechanical pulping. Due to its flammability, paperboard should not be used in high-temperature applications.
Forming	Compatible with common paper forming and assembly processes, such as various cutting processes and folding. Joining with water-based adhesives doesn't affect the recyclability and biodegradability of the material.
Finishing	Barrier paperboard materials are available in both bleached (white) and brown, unbleached versions. Both alternatives can be printed, but certain materials may be optimized for specific printing processes and inks. Low-migration, petrochemical-free inks should be used to improve the recyclability and biodegradability of the paperboard.

Barrier paperboard packaging by Grönska, featuring a print design by Saga Maria Sandberg. The packaging is biodegradable and can be composted with paper and other biodegradable waste.

BASILIKA
Närodlad
Komposterbar kruka
Obesprutad
HÄRODLADE GRÖDOR
GRÖNSKA
VERTIKAL ODLING

Alternative-fibre paperboard

Like alternative-fibre packaging paper, alternative-fibre paperboard has the potential to reduce agricultural waste and the emissions from stubble burning, a major source of pollution.

Suppliers & materials	- PaperWise straw-based paperboard - PropalPack bagasse paperboard	
Raw material origin	PaperWise works with farmers and local paper mills in India to recover agricultural waste such as stubble, which would otherwise likely be burnt, causing severe local pollution. PropalPack uses bagasse, the leftover waste after sugarcane has been crushed, from farmers in Colombia to produce their packaging papers.	
	PaperWise straw-based paperboard[10]	**FSC paperboard**[11]
GWP	0.65 kg CO_2e / kg	1.2 kg CO_2e / kg
Energy use	No data	18.5 MJ / kg
Water use	No data	2,490 l / kg
Toxicity	Paper is considered non-toxic, but the suitability of specific paper materials for food-contact applications should be confirmed with suppliers, as it can react in non-toxic but unexpected ways, such as altering the colour or taste of foods. None of the alternative-fibre paperboard materials in this book use elemental chlorine for bleaching.	
Circularity	Recyclable with other paper waste, so long as the material is kept dry and free from excessive food waste. Contaminated paper that cannot be recycled can be composted at home or in industrial composting facilities instead. Printing inks, water-based adhesives and recyclable paper-based labels do not affect the recyclability or biodegradability of the paper. For an overview of recycling rates and circular design guidelines for paper, see page 191.	
Mechanical properties	Tough and rigid, paperboard materials also offer good compression strength. Request a technical data sheet for specific materials, weights and grades.	
Environmental resistance	Paperboard without a barrier coating has poor resistance to moisture and grease. Paper that is made using chemical pulping - also known as kraft paper - has better UV resistance than paper made using mechanical pulping. Due to its flammability, paper should not be used in high-temperature applications.	
Forming	Compatible with common paper forming and assembly processes, such as various cutting processes and folding. Joining with water-based adhesives does not affect the recyclability and biodegradability of the material.	
Finishing	Available in bleached (white) and brown, unbleached versions. Some alternative-fibre paperboard materials have visible fibre and particles in the surface. The materials are printable, but certain materials may be optimized for specific printing processes and inks. Low-migration, petrochemical-free inks should be used to improve the recyclability and biodegradability of the material.	

Squalane Oil by Biossance. The packaging uses paperboard made with sugarcane waste, also known as bagasse.

BIOSSANCE
100%
SQUALANE OIL
Sugarcane-Derived Oil to Hydrate Body, Face, & Hair
Huile dérivée de la canne à sucre pour hydrater le corps, le visage et les cheveux
PLANT DERIVED SQUALANE
WHAT IS SQUALANE (C30H62)?
BIOSSANCE

Moulded paper pulp

The latest generation of moulded paper pulp materials offers high-quality, smooth surfaces that can replace plastic moulded parts in many applications, while offering all the environmental benefits of conventional paper materials.

Suppliers & materials	- PaperFoam® moulded paper pulp - Knoll Ecoform™ moulded paper pulp - Billerud FibreForm® moulded paper pulp
Raw material origin	Responsibly sourced timber from managed forests. Ask suppliers to confirm relevant certification, including FSC, PEFC and EU Ecolabel. PaperFoam is a two-component material that consists of cellulose fibre and industrial starch made with potatoes.
	PaperFoam® moulded paper pulp[12]
GWP	0.9 kg CO_2e / kg
Energy use	11.6 MJ / kg
Water use	2.2 l / kg
Toxicity	Pure moulded paper pulp is considered non-toxic, but the suitability of specific materials for food-contact applications should be confirmed with suppliers, as it can react in non-toxic but unexpected ways, such as altering the colour or taste of foods. None of the moulded paper pulp materials in this book use elemental chlorine for bleaching.
Circularity	Recyclable with other paper waste, so long as the material is kept dry and free from excessive food waste. Contaminated paper that cannot be recycled can be composted at home or in industrial composting facilities instead. Printing inks, water-based adhesives and recyclable paper-based labels do not affect the recyclability or biodegradability of paper. For an overview of recycling rates and circular design guidelines for paper, see page 191.
Mechanical properties	Moulded paper pulp offers decent dimensional stability and compression strength, as well as good energy absorption. Part strength and rigidity can be improved by adding ribs, selectively thicker wall sections and other features.
Environmental resistance	Poor resistance to moisture and grease. Moulded paper pulp should not be used in high-temperature applications due to its flammability.
Forming	The specific moulding processes differ between the materials featured here, but all have in common that they are variations of compression moulding, or, in the case of PaperFoam®, injection moulding.
Finishing	Compared to conventional recycled moulded paper pulp, the materials featured here offer significantly higher-quality surfaces that allow for detailed printing and fine surface textures. The material can be dyed for solid colour applications. Low-migration, petrochemical-free inks and dyes should be used to improve the recyclability and biodegradability of the moulded paper pulp.

PaperFoam® samples. This moulded paper foam material is made of a mix of cellulose fibre and potato starch that can be injection moulded for complex and highly detailed parts.

we
care
mit unseren erneuerten
TOP-Smartphones

7
Emerging sustainable material technologies

While most of the materials and technologies presented in this book are fairly well established, commercially available, and can be seen as a logical progression of the conventional material industries, a growing number of materials suppliers, as well as scientists, artists and designers, are exploring emerging sustainable material technologies that don't fit so easily into traditional categories.

The work of Teresa van Dongen illustrates perfectly this willingness to explore and experiment with alternative technologies without preconceptions. Teresa's Aireal materials library (**aireal-materials.com**) maps out a selection of materials derived from carbon capture and utilization (CCU) technologies, giving an excellent introduction to this fascinating and very active area of materials science. You can read more about how her work cuts through science and design on pages 218 to 221.

Material suppliers and scientists are also looking increasingly to natural growth processes for inspiration. One example is mycelium, the fine network of roots that connect fungi. In the context of design, mycelium that has simply been grown into shape is being used as an alternative to plastic foam in packaging, for example. It can also be further processed to create other materials such as artificial leather. For more on mycelium and other examples of natural growth processes, see pages 224 to 225.

Lastly, we perhaps need to take a closer look at our conventional, fairly linear approach to recycling. Is it reasonable to expect pure, uncontaminated recycled materials when waste streams tend to be so complex? Chemical recycling of plastics has huge potential for a new generation of high-quality recycled materials, but maybe we should also consider new blends and hybrids as a legitimate alternative going forward. Or perhaps we should change the way we make materials in the first place, avoiding composites and complex material combinations altogether. The last section of this book investigates all of these approaches – chemical recycling (pages 226–27), mixed-plastics recycling (pages 228–29) and mono-material plastic composites (pages 230–31).

Emerging technologies

An overview of the emerging sustainable material technologies featured in this chapter.

Carbon capture and utilization (CCU; p.222)

Conversion
Using CO_2 to produce chemicals and other raw materials

- LanzaTech
- Fairbrics

Direct use
Using CO_2 without conversion as a raw material in its own right, or using materials that simply absorb CO_2

- Carbstone
- GreenSand
- Green Minerals

Natural growth processes (p.224)

Microorganisms
Harnessing the role of bacteria and other microorganisms in natural processes such as fermentation and the formation of corals

- SCOBY
- BioBasedTiles®

Mycelium
Mycelium, the fine network of roots that connects fungi, is becoming a useful raw material in product design

Alternative approaches to plastic recycling

New recycling processes

Chemical recycling (p.226)
Breaking plastic waste down into its chemical building blocks

- Dissolution
- Polymerization
- Conversion

Mixed-plastics mechanical recycling (p.228)
Recycling different plastic materials without separation

- OmniPO
- SABIC XENOY™ PC/PET
- EcoRub

Mono-material plastic composites (p.230)
Composites entirely made from a single plastic material, recyclable in normal plastic waste streams

- Biaxially oriented film
- Self-reinforced composites

'I'm interested in the role that design can play in the scientific development of new materials.'

An interview with Amsterdam-based designer Teresa van Dongen, whose work explores the intersection between design and biology through such topics as bioluminescence, electroactive bacteria and the use of CO_2 as a raw material. Find out more about her work at teresavandongen.com, and about Aireal – a selection of CO_2-based material technologies she has curated – at aireal-materials.com

I read an interview where you talked about going for a walk in the forest and finding a mud well with the strongest ecosystem of electroactive bacteria you'd ever encountered. This doesn't sound like the average designer out on a walk.

I'd studied biology for a couple of years before I applied to the Design Academy in Eindhoven here in the Netherlands. I didn't have a clear plan to combine these two subjects, but after a while at the academy I started to experiment with fungi, and it suddenly became clear to me that design has a role to play in the scientific development of microbial- and bio-based materials, as well as other processes and technologies that I'd come across during my time as a biology student. One thing that really stuck with me from that time was the work that one of my biology professors was doing to get algae to produce biofuels using natural processes. When I was graduating I contacted him to ask what had happened to this research and found out that while he had solved many difficult technical aspects of the process, at that point there were no clear applications for the technology in everyday life. Inspired by his work, I decided to explore this cross-section between design and biology to find out what these applications might look like – such as Ambio, a lamp I designed with bioluminescent bacteria that light up when you give the lamp a gentle push.

Mud Well installation by Teresa van Dongen,
Terschelling island, the Netherlands, 2019.

Your ability to harness the power of biology has the potential to change how we think about certain processes and materials. For most of us, carbon dioxide and other greenhouse gases are just pollution, but with your work on Aireal you're presenting an alternative view.

I was working on a project at Ghent University in Belgium when the researchers at the Center for Microbial Ecology and Technology there showed me their experiments with converting greenhouse gases into useful things like nutrients and materials. For my own work, I wanted to look at CO_2. In the context of fossil fuels and industry, CO_2 is pollution, but it's also an irreplaceable part of the cycle of life and a crucial building block of all plant life on the planet. Rather than simply seeing it as waste, I wanted to look for processes that use CO_2 as a raw material and I ended up unearthing a network of different initiatives, research groups and universities exploring this field. The amount of work being done was overwhelming, so I decided I needed to collect information and create some kind of selection to make sense of it all. That was the starting point for Aireal, a library I created for materials that are made with CO_2.

Once I had this platform, certain patterns started to emerge. For example, there are materials in the collection that embody natural processes taking place around us all the time, such as olivine – a very common volcanic mineral that absorbs CO_2. If olivine is ground up into sand, it can absorb more CO_2 because the reaction will occur over a bigger surface area. One tonne of olivine sand can absorb about a tonne of CO_2 under natural conditions. It can be used anywhere that sand or gravel are useful, such as on train tracks. Also, once the material has fully reacted with CO_2 its properties have changed, making it a very useful raw material for material production. The only drawback is that the process is rather slow; olivine sand takes up to ten years to absorb the maximum amount of CO_2 that it can hold. But this can be reduced to just a few minutes inside a reactor, resulting in a mineral powder that can be used for paper, cement, or any other application where similar mineral powders are used. Olivine can only take up half of its mass in CO_2 in a reactor, which is less than under natural conditions, but the controlled conditions make it much easier to harvest for future material use.

The use of reactors and other lab-based processes leads to the more science-based and fabricated material technologies in the Aireal collection. Cardyon® is a partially CO_2-based polyol from the German plastics supplier Covestro. Polyols are a key ingredient in soft plastics like polyurethane foam and thermoplastic polyurethane (TPU) elastomers, where the use of CO_2 as a raw material could be a direct replacement for petrochemical-based chemicals. There is also Carbstone, a replacement for concrete that uses waste from the iron mining industry as a raw material and CO_2 as the hardening agent (see page 222). Carbstone is capable of absorbing roughly the amount of CO_2 that's emitted when producing conventional concrete, potentially reversing the carbon footprint of the concrete industry, and reducing waste from the iron and steel industry in the process.

All the technologies in the Aireal library are at various stages of development and the challenge now is for these materials to find their way into real-world applications. In some cases, it may be difficult to find commercial incentives – after all, it's not possible to patent olivine sand – but this is where design can have a role to play.

Materials from the Aireal collection, left to right, from top

Row 1: olivine concrete (Green Minerals); microbial proteins (Avecom)

Row 2: olivine reacted mineral powder; Cardyon® flexible foam (Covestro, Carbon4PUR)

Row 3: PUR hard foam (Covestro, Carbon4PUR); polyurethane elastic thread (Covestro, RWTH Aachen Institute of Textile Technology, Carbon4PUR)

Row 4: olivine paper (Green Minerals); raw ground olivine mineral

Carbon capture and utilization (CCU)

Most of us tend to think of CO_2 simply as pollution, but as the interview with Teresa van Dongen shows (see pages 218–21), this is only part of the story. CO_2 is also a prerequisite for life on this planet and a key raw material in many natural processes, such as the growth of plants. Several processes based on so-called carbon capture and utilization (CCU) are now emerging, with the aim of using carbon oxides like CO and CO_2 for conversion into various fuels and chemicals as well as a raw material in their own right.

Some CCU technologies can be used by steel mills, power plants and other energy-intensive manufacturing sites to capture carbon at the source, preventing it from being released into the atmosphere. US-based LanzaTech is one example, whose CCU installations to date include a pilot plant at a steel mill in China's Hebei province. LanzaTech can then convert this carbon into many different types of chemicals, such as the raw materials needed to produce the EVA foam used in the sole of On Running's CleanCloud™ trainer. Additionally, the upper in the shoe uses CCU-based polyester textiles from the French supplier Fairbrics. According to Fairbrics, their textiles not only remove carbon from the atmosphere, they also use less energy than conventional petrochemical-based polyester.

Other approaches to CCU make use of captured carbon without conversion. The Belgian green tech companies VITO and Orbix use waste from the steel industry as the raw material and captured carbon as the hardening agent when making Carbstone, a CCU-based alternative to concrete. The amount of carbon sequestered in each cubic metre of Carbstone is roughly equal to the amount of CO_2 emitted during the production of a cubic metre of conventional concrete, so the process has huge potential for reversing the massive environmental impact of the concrete industry.

A third approach is to simply let natural carbon capture processes run their course. Olivine, an abundant volcanic mineral, naturally absorbs CO_2 from the atmosphere, and when olivine rocks are ground up into sand or gravel, the mineral's capacity for capturing carbon grows exponentially, due to the larger surface area. According to GreenSand, a Netherlands-based olivine sand supplier, one tonne of olivine sand is capable of absorbing one tonne of CO_2 from the atmosphere. GreenSand has launched several projects in the Netherlands, including the gravel bed for the tracks of a trainline between Rotterdam and Hook of Holland. Another Dutch company, Green Minerals, has developed a process for speeding up the CO_2 absorption process of olivine in a reactor, making it possible to produce olivine powder fully charged with CO_2 that can be used in applications where similar minerals are used, such as coated paper, cement and functional fillers for plastic materials.

CleanCloud™ CloudPrime running shoe by On Running. The shoe uses two materials derived from carbon capture technologies – the upper is made with a polyester textile from Fairbrics, and the EVA foam midsole was developed in collaboration with LanzaTech.

Carbinox slab by VITO and Orbix. Carbinox is a concrete-like material made with slag waste from the steel industry, using a process that absorbs roughly the same amount of CO_2 as is emitted during the production of an equal volume of conventional concrete.

LanzaTech's carbon capture installation at a Shougang Group steel mill in Hebei, China.

Natural growth processes

Many strong, durable materials and structures found in nature are the result of natural growth processes. An increasing number of suppliers are developing techniques for 'taming' these processes so that they can be used for growing parts and components for product design.

Probably one of the best-known examples of harnessing the power of natural growth processes, mycelium is the fine network of roots that connects fungi, and a fascinating topic in itself – for an introduction, have a look at the amazing book *Entangled Life* by Merlin Sheldrake. In terms of design, several companies have developed ways for turning these structures of interconnected microorganisms into surprisingly strong and tear-resistant materials that are also soft to the touch and naturally water-repellent. The Vancouver-based personal care brand Well Kept, for example, use mycelium for the tray insert in their razor packaging, rather than plastic foam or a vacuum-formed tray.

SCOBY, or 'symbiotic culture of bacteria and yeast', to give its full name, is also a material based on natural growth – specifically, bacteria. This jelly-like biofilm is a by-product of the fermentation process that occurs naturally in foods and drinks like kimchi, soy sauce and kombucha. When it's dried, it becomes a strong and durable yet flexible material with similar properties to leather. For the time being, most SCOBY leather – or kombucha leather, as it's also known – is made in small volumes by designers such as Studio Lionne van Deursen, demonstrating the usefulness of the material in a wide range of applications, including interior accessories, jewellery and experiments like the Unfold project featured opposite.

The growth and formation of corals is another example of a natural process that can be harnessed for the purposes of design. By studying the interactions between the different microorganisms that combine to form corals, Biomason has developed a process for growing Biocement® – an alternative to conventional Portland cement that can be grown at ambient temperatures and without CO_2 emissions. The first commercially available product from Biomason is tiles made from recycled granite with Biocement® as the binder. These are known as Biolith® in the US and BioBasedTiles® in Europe.

BioBasedTiles® by StoneCycling and Biomason. The tiles are grown using a process based on the way that corals form in nature.

Studio Lionne van Deursen's Unfold project uses origami folding techniques to explore the functionality of SCOBY sheet materials.

Shaver packaging by Well Kept. The box insert is made of mycelium that is grown into shape.

Plastic chemical recycling

In the context of plastics, chemical recycling refers to recycling processes that are capable of breaking down plastic waste into its chemical building blocks, which can then be used to make recycled plastics or other chemicals altogether.

Unlike mechanical recycling of plastics, chemical recycling is capable of producing recycled plastic materials with essentially identical properties to pure, virgin materials – any pigments, additives and other contaminants in the plastic waste are separated out during the process. In some cases it's even possible to recycle different types of plastic together without separation. Despite these huge benefits compared to mechanical recycling, and the fact that it becomes possible to recycle a wider range of plastic waste, it should be noted that chemical recycling can be rather energy-intensive.

Broadly speaking, there are currently three types of chemical recycling. To get the most out of them, it helps to know a little bit about how plastics are made in the first place. Most plastics are made with hydrocarbon feedstocks that are further refined into monomers, then finally a chain reaction turns monomers into polymers, or plastics.

The first kind of chemical recycling is called dissolution. This removes any pigments, additives and other contaminants from plastic waste, turning it back into a pure polymer. It requires plastic waste to be sorted by type, but it can be a relatively efficient process; the US-based supplier PureCycle reports that CO_2e emissions from recycled polypropylene using their proprietary dissolution process are roughly 35 per cent lower than from virgin petrochemical-based polypropylene.

The second type of chemical recycling is depolymerization. This also removes any contaminants but goes a step further by breaking plastic waste down into monomers that can then be further refined into recycled polymers. This offers greater flexibility as the same type of monomer can be used to produce several different types of plastic. Eastman, a plastics supplier with a large portfolio of copolyester materials, has developed a depolymerization process that uses monomers from chemically recycled PET waste to produce two copolyester materials: Tritan Renew and Cristal Renew.

The third chemical recycling process, which breaks plastic waste down into feedstocks, is known as conversion. Unlike dissolution and depolymerization, conversion processes can typically take in mixed plastic waste. London-based Plastic Energy have built two chemical recycling plants that are providing plastic suppliers with TACOIL™, a feedstock that can be used in the production of ethylene and propylene, both of which are key ingredients in a wide range of plastic materials. CO_2e emissions for their low-density polyethylene (LDPE), produced with TACOIL™, are 55 per cent lower than those from virgin material.

The European Chemical Industry Council (CEFIC) provides an excellent overview of chemical recycling processes, key suppliers and case studies at **cefic.org**.

Key types of chemical recycling

Dissolution removes contaminants from plastic waste, resulting in high-quality recycled plastic with similar properties to virgin material. Depolymerization removes any contamination and also breaks plastic waste down into monomers to produce various recycled plastics. Conversion turns plastic waste into feedstocks that, in principle, can be used to make any kind of recycled plastic.

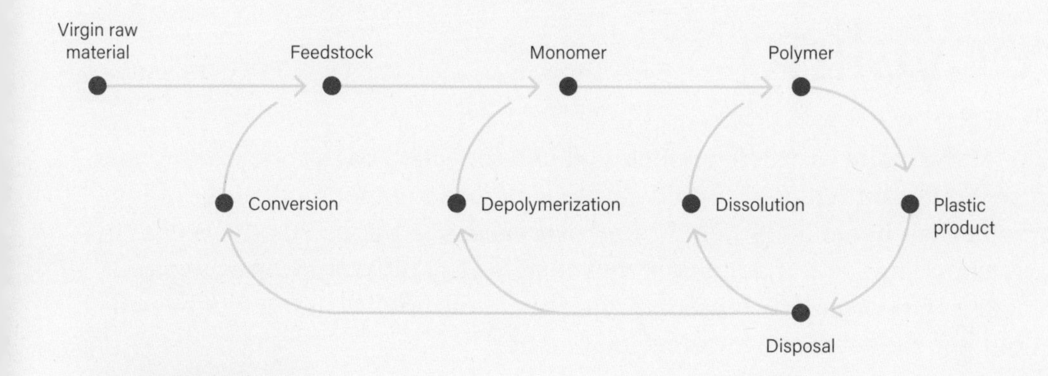

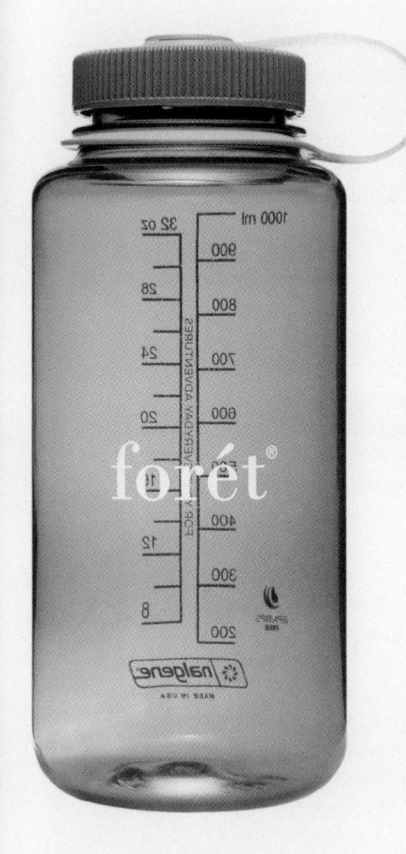

A special edition of the classic Nalgene bottle designed by Forét®. The bottle is made with Tritan Renew, a copolyester material from Eastman, partially made with recycled PET bottles using depolymerization.

Plastic Energy's chemical recycling plant in Almeria, Spain. Plastic Energy is currently operating two chemical recycling plants, providing TACOIL™ recycled feedstock to suppliers for further processing into various plastic materials.

Mixed-plastics mechanical recycling

One of the major obstacles holding plastic recycling back today is the need to sort different types of plastic carefully before the mechanical recycling process can begin in earnest (for an overview, see pages 20–22). While this only underlines that mechanical recycling is very different from making plastics with virgin raw materials, we should perhaps also think differently about our expectations for recycled plastics.

Omni Polymers is a Swedish recycler with a special focus on recycling flexible plastic packaging. This type of packaging often consists of several layers of different plastics and other materials for enhanced performance, such as improved liquid and oxygen sealing, or antistatic properties, to give just a few examples. These layers are difficult, if not impossible, to separate for recycling. In response, Omni Polymers have developed a process for mechanically recycling these materials without separation, resulting in a blend of plastics that consists mostly of polypropylene and polyethylene. At the time of writing, actual material performance data is hard to come by, but Omni Polymers report that several customers have run successful tests with the material, and a pilot plant sponsored by Nestlé and the Swedish Environmental Protection Agency was opened in October 2022 in southern Sweden, with the capacity to produce 15,000 tonnes of recycled material annually.

Beyond packaging, the global plastics supplier SABIC has developed a polycarbonate material which contains 20 per cent post-consumer recycled PET. According to SABIC, the PET waste is collected from the open environment, meaning that the material is often severely degraded by exposure to UV radiation, sea water and the elements in general. By blending the recycled PET material with virgin polycarbonate, SABIC is able to offer a material with sufficiently high performance to go into such demanding applications as Microsoft's Ocean Plastic mouse.

EcoRub, another Swedish plastics recycler, specializes in elastomers, a group of soft and elastic plastics that are currently not widely recycled. EcoRub's TPRR® (thermoplastic recycled rubber) material is a blend of post-consumer recycled tyre rubber and polypropylene that can be formed using common thermoplastic processes such as injection moulding and extrusion. This opens up a wide range of new applications, as pure thermoset rubber is incompatible with these processes.

While it could be argued that the difficulties of recycling products that use mixed plastic materials is only compounded by recyclers that introduce more mixed plastics into the market, this has to be balanced by the fact that millions of tonnes of plastic waste are routinely burnt and put into landfill, or worse, end up in the open environment. In this context, new recycling processes that keep plastic waste in circulation for longer could have real value.

EcoRub TPRR® samples. The material is made with a blend of PCR tyre rubber and polypropylene that can be formed using thermoplastic processes such as injection moulding and extrusion.

Ocean Plastic mouse by Microsoft. The exterior is made with XENOY™, a blend of virgin polycarbonate and PCR PET developed by Sabic. The recycled PET material is clearly visible in the coloured specks in the surface of the product.

Mono-material plastic composites

Plastic composites can offer greatly enhanced properties and functionality, but because they are made of blends of different materials they can also be difficult to recycle. Rather than trying to find new recycling solutions for composites, a growing number of suppliers are developing monomaterial plastic composites, entirely made with a single material.

As mentioned in the entry on mixed plastics recycling, flexible packaging film is often made with a combination of different materials for improved sealing and other properties. The issue with using film made with a single plastic in packaging applications that require good sealing properties is that the strands of molecules that make up the material tend to be oriented in the same direction, which allows oxygen and liquids to pass through. Taking a closer look at how plastic films are made in the first place, suppliers are developing processes for manufacturing what is known as biaxially oriented film, which has much better sealing properties in a single material. Biaxially oriented polyethylene (BOPE), polyropropylene (BOPP) and PET (BOPET) films are available from several suppliers, including Ampacet, Mitsubishi and SABIC.

Rigid plastic parts in demanding applications that require enhanced strength and impact resistance – for example, in the automotive industry, sports equipment and aerospace applications – are often made with composites that consist of a plastic matrix, or binder, mixed with a reinforcing textile or loose textile fibres. The downside is that only specialist recyclers are able to process these materials, and even then there are several disadvantages. Because mechanical recycling involves grinding up the composite waste, any textiles used will be shredded, greatly diminishing their properties, since reinforcement, and the length of any loose fibres used, will be shortened with each cycle, which also reduces strength.

An alternative approach is to create composite structures from a single material: so-called self-reinforced composites. Several polypropylene-based examples are commercially available, including CURV®, from Propex Furnishing Solutions; Torodon®, from Don & Low; and PURE®, from DIT Weaving. These are all sheet materials that consist of conventional polypropylene fibre that adds strength and rigidity to the material. A handful of specialist suppliers, including Comfil in Denmark, offer self-reinforced composites made with other materials, such as PET and PLA. These materials can be recycled with other plastic waste of the same type, while offering performance that can compete with conventional composites in many cases.

Coffee bag by Coffee Collective. The bag is made entirely with biaxially oriented polyethylene (BOPE) film, with an exterior recycled layer and an interior layer made with virgin material. The bags can be recycled with other polyethylene waste without separation.

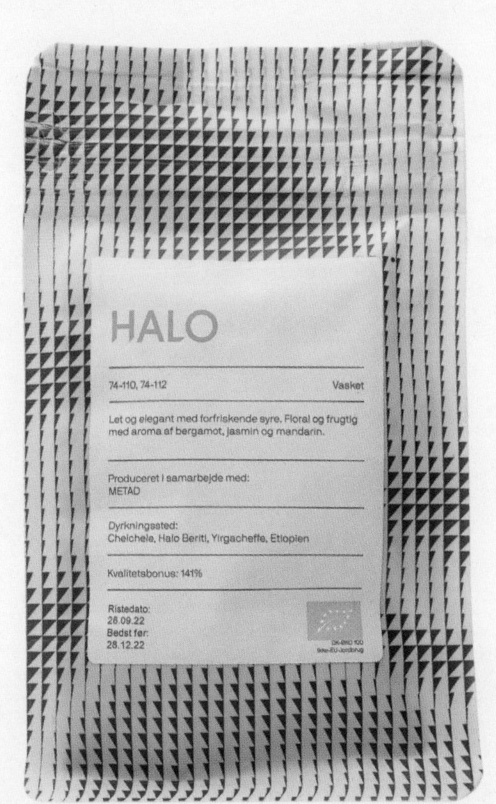

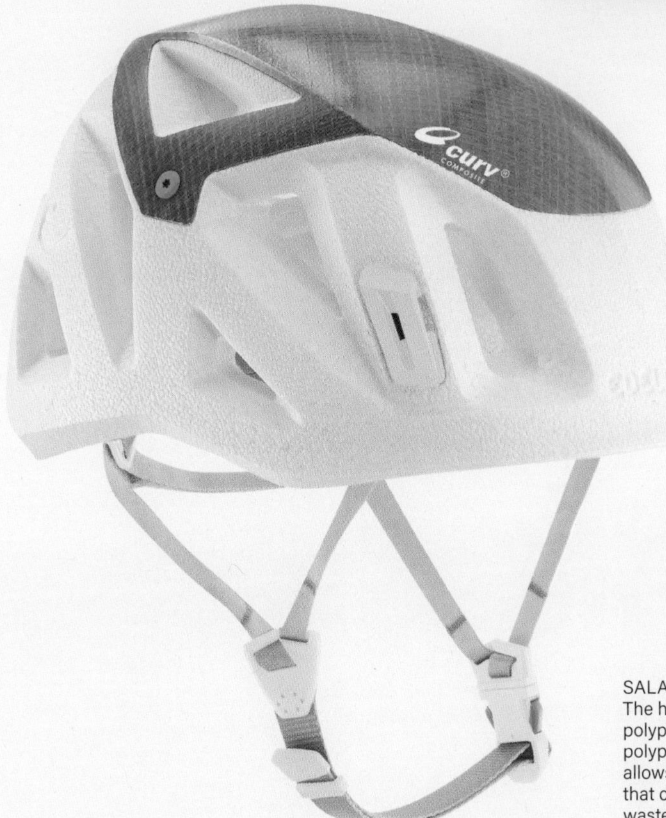

SALATHE LITE climbing helmet by Edelrid. The hard shell is made with Curv® self-reinforced polypropylene composite, with an expanded polypropylene foam core. This combination allows for a lightweight monomaterial assembly that can be recycled with other polypropylene waste without separation.

Notes

Introduction

1. Global recycling rates: steel (*World Steel in Figures*, 2021, and 'Scrap Use in the Steel Industry' fact sheet, 2021 – both World Steel Association); stainless steel (*Stainless Steel in Figures*, 2022, and 'The Global Life Cycle of Stainless Steels', 2016 – both WorldStainless); aluminium (annual global statistics, 2020, and 'Aluminium Recycling Factsheet', 2020 – both International Aluminium Institute); paper (Bureau of International Recycling, 'Paper and Board Recycling in 2019'); plastics (Organisation for Economic Co-operation and Development, *Global Plastics Outlook*, 2020); glass (Dr.-Ing. J. Harder, OneStone Consulting, 'Glass Recycling – Current Market Trends', *Recovery* magazine, May 2018); textiles (Textile Exchange, *Preferred Fiber & Materials Market Report*, 2021)

Chapter 1: Plastics

Plastic recycling

1. Plastics Europe, *The Circular Economy for Plastics*, 2022
2. Plastics Recyclers Europe, *Report on Plastics Recycling Statistics 2020*
3. Eunomia for Plastics Recyclers Europe, *PET Market in Europe: State of Play*, 2022; *HDPE and PP Market in Europe: State of Play*, 2020; and *Flexible Films Market in Europe: State of Play*, 2020
4. Fortum Circo® LCA
5. Plastics Europe, Eco-profiles
6. Eunomia for Plastics Recyclers Europe, *HDPE and PP Market in Europe: State of Play*, 2020
7. Fortum Circo® LCA
8. Plastics Europe, Eco-profiles
9. Eunomia for Plastics Recyclers Europe, *HDPE and PP Market in Europe: State of Play*, 2020
10. Indorama Deja™ PCR PET report
11. Plastics Europe, Eco-profiles
12. Eunomia for Plastics Recyclers Europe, *PET Market in Europe: State of Play*, 2022
13. MBA Polymers Carbon Trust certification
14. Plastics Europe, Eco-profiles
15. Covestro Circular CMF guide
16. Plastics Europe, Eco-profiles
17. Aquafil ECONYL® PCR PA6 EPD
18. Plastics Europe, Eco-profiles
19. TRINSEO™ APILON™ 52 Bio LCA, 2021
20. European TPU average, as estimated by TRINSEO™ APILON™ 52 Bio LCA, 2021
21. Author's estimate, based on IDEMAT database
22. European Commission, JRC Environmental Footprint database
23. Idemat 2022 database
24. Idemat 2022 database; EC, JRC Environmental Footprint database
25. European Tyre and Rubber Manufacturers' Association (ETRM) press release, 11 May 2021

Renewable plastics

1. European Bioplastics, *Bioplastics Market Development Update 2021*
2. Ibid.
3. Borealis LCA
4. Plastics Europe, Eco-profiles
5. Borealis Bornewables™ LCA
6. Plastics Europe, Eco-profiles
7. Total Corbion Luminy® PLA
8. Idemat 2022 database
9. SABIC LCA
10. Plastics Europe, Eco-profiles
11. DSM EcoPaXX® LCA
12. Plastics Europe, Eco-profiles
13. Spolchemie ENVIPOXY® EPD
14. European average, EC, JRC Environmental Footprints database
15. TRINSEO™ APILON™ 52 BIO LCA
16. European TPU average, as estimated by TRINSEO™
17. EC, JRC Environmental Footprints database
18. ETRMA press release, 11 May 2021

Chapter 2: Textiles

1. Data from: Textile Exchange, *Preferred Fiber and Materials Market Report*, 2021

Textile recycling

1. Textile Exchange, 2021
2. Teijin ECOPET® PCR PET LCA; N. M. van der Velden et al., 'LCA Benchmarking Study on Textiles Made of Cotton, Polyester, Nylon, Acryl, or Elastane', *International Journal of Life Cycle Assessment*, 2012
3. Van der Velden et al., 2012
4. Textile Exchange, 2021
5. Aquafil ECONYL® PCR PA6 yarn EPD; Van der Velden et al., 2012
6. Van der Velden et al., 2012
7. Textile Exchange, 2021
8. Fibre: IDEMAT 2023 database; woven textile: Van der Velden et al., 2012
9. SCS Global Services, 2017
10. Textile Exchange, 2021
11. Fibre: F. A. Esteve-Turrillas and M. de la Guardia, 'Environmental Impact of Recover Cotton in Textile Industry', *Resources, Conservation and Recycling*, January 2017; Textile weaving: Van der Velden et al., 2012
12. PE International for Textile Exchange, *The Life Cycle Assessment of Organic Cotton Fiber – A Global Average*, 2014; Textile weaving: Van der Velden et al., 2012
13. Textile Exchange, 2021
14. Manteco MWool® LCA
15. S. G. Wiedemann et al., 'Environmental Impacts Associated with the Production, Use, and End-of-life of a Woollen Garment', *International Journal of Life Cycle Assessment*, 2020
16. Textile Exchange, 2021
17. Author's estimate based on average reduced impact of recycled leather provided by ELeather
18. Product Environmental Footprint (PEF) database
19. Textile Exchange, 2021

Renewable textiles

1. Virent renewable paraxylene LCA
2. Van der Velden et al., 2012
3. Fulgar EVO LCA; Van der Velden et al., 2012
4. Van der Velden et al., 2012
5. Fibre: IDEMAT 2023 database; woven textile: Van der Velden et al., 2012
6. Textile Exchange, 2021
7. SCS Global Services, 2017
8. Textile Exchange, 2021
9. Modern Meadow Bioleather1 LCA
10. PEF database
11. PE International for Textile Exchange, 2014; textile weaving: Van der Velden et al., 2012
12. Ibid.

13. Textile Exchange, 2021
14. Fibre: Wiedemann et al., 2020. Median value. Wool fibre LCA GWP data varies between 10.4 kg CO_2e / kg and 103 kg / kg. Textile weaving: Van der Velden et al., 2012
15. PEF database
16. Textile Exchange, 2021

Chapter 3: Metals

1. Metals emissions: steel (*Steel Facts*, 2018, and *World Steel in Figures*, 2021 – both World Steel Association); aluminium (International Aluminium Institute [IAI], *Greenhouse Gas Emissions – Primary Aluminium*, 2018)
2. Ibid. For sources on individual steel and aluminium materials, see book profiles

Metal recycling

1. Steel (*Steel Facts*, 2018, and *World Steel in Figures*, 2021 – both World Steel Association); aluminium (*Annual Global Statistics*, 2020, and Recycling Fact Sheet, 2020 – both IAI); stainless steel (WorldStainless's *Stainless Steel in Figures*, 2022, and *The Global Life Cycle of Stainless Steels*, 2016)
2. Hydro CIRCAL 6000-series aluminium ingot EPD
3. Product Environmental Footprint (PEF) database, aluminium ingot (silicon and magnesium main solutes), EU average
4. Novelis HRC57S® 5005 alloy, estimated by Novelis
5. PEF database, aluminium ingot (magnesium main solute), EU average
6. Raffmetal SILVAL aluminium ingot EPD
7. PEF database, aluminium ingot (silicon main solutes), EU average
8. ThyssenKrupp Bluemint® Recycled hot rolled coil, DIN EN ISO/IEC 17029 and TÜV SÜD VERIsteel statement of conformity
9. World Steel Association hot rolled coil LCI study, 2020
10. Outokumpu stainless steel cold rolled coil EPD
11. PEF database, stainless steel cold rolled coil world average

Low-carbon metals

1. Hydro REDUXA 6000-series aluminium ingot EPD
2. PEF database, aluminium ingot (silicon and magnesium main solutes), EU average
3. ThyssenKrupp Bluemint® Pure hot rolled coil DNV Independent Limited Assurance Report
4. World Steel Association, 2020 hot rolled coil LCI study, 2020

Chapter 4: Ceramics and glass

Ceramic recycling

1. European Environment Agency (EEA), 'Construction and Demolition Waste: Challenges and Opportunities in a Circular Economy', 2020
2. Eurostat, 'Mineral Waste from Construction and Demolition, Waste Treatment', 2020; EEA, 'Construction and Demolition Waste: Challenges and Opportunities in a Circular Economy', 2020
3. StoneCycling EPD, Salami WasteBasedBrick®
4. Product Environmental Footprint (PEF) database, EU average
5. Herrljunga Terrazzo HT-LYKKE EPD
6. Cosentino Dekton EPD
7. ÖKOBAUDAT database, 20-mm marble slab

Glass recycling

1. Packaging glass and flat glass, world (Dr.-Ing. J. Harder, OneStone Consulting, 'Glass Recycling – Current Market Trends', *Recovery* magazine, May 2018); packaging glass, EU (European Container Glass Federation, 'Year 2018 Glass Recycling Statistics'); flat glass, EU (European Federation of Glass Recyclers, 'Statistics 2014–2018')
2. PEF database, 80% recycled packaging glass, EU and European Free Trade Association (EFTA) average
3. PEF database, virgin packaging glass, EU and EFTA average
4. Katy Devlin, 'Flat Glass Recycling', *Glass Magazine*, 22 March 2022
5. Saint-Gobain DIAMANT 3-mm 30% recycled flat glass EPD
6. ÖKOBAUDAT database, 4-mm virgin flat glass
7. MAGNA Glaskeramik® EPD
8. Idemat database, estimate
9. Ibid.

Chapter 5: Wood

1. Global forest-related greenhouse gas fluxes, Global Forest Watch, tinyurl.com/mr2wsrbe (accessed Jan. 2023)
2. Global forest area (Statistics Division of the Food and Agriculture Organization of the United Nations [FAOSTAT], global forest area 2015–2020); certified forest area (Food and Agriculture Organization of the United Nations [FAO] and United Nations Economic Commission for Europe, *Forest Products Annual Market Review 2020–2021*, graph 1.4, FSC- and PEFC-certified forest area, 2013–2020)
3. Global timber trade in 2020, FAOSTAT, Forestry Production and Trade, 2020, tinyurl.com/47hseb7m. (accessed Jan. 2023)
4. J. Bojesen Jensen, 'An Investigation into the Suitablility of Pawlonia as an Agroforestry Species for UK & NW European Farming Systems', Coventry University, 2016
5. Global PIR and PCR wood waste in 2020: FAOSTAT, Forestry Production and Trade, 2020, tinyurl.com/47hseb7m. (accessed Jan. 2023). Recovered PCR wood converted from tonnes to m^3, based on the assumption that the average weight of 1 m^3 of wood is 1.5 tonnes
6. Panguaneta PureGlue™ NAF plywood EPD
7. FINSA Fibracolour NAF MDF EPD
8. MEDITE® SMARTPLY NAF OSB EPD
9. Abet Laminati PRINT HPL Thin EPD
10. Amorim expanded corkboard EPD

Chapter 6: Paper

Paper recycling

1. Bureau of International Recycling (BIR), *Paper and Board Recycling in 2019*
2. Ibid.
3. Ibid.
4. ArjoWiggins Cyclus Pack paper profile
5. Product Environmental Footprint (PEF) database, EU average
6. ArjoWiggins Teknocard paper profile
7. PEF database
8. Idemat 2022 database

Virgin paper

1. Statistics Division of the Food and Agriculture Organization of the United Nations (FAOSTAT), Forestry Production and Trade database, 2020
2. Ibid., plus FAO, *Pulp and Paper Capacities, 2020–2025*
3. Arctic Paper Munken Kraft paper profile
4. UPM Kymi mill paper profile
5. UPM Jämsänkoski mill paper profile
6. PaperWise LCA study
7. PEF database, EU average
8. Billerud CrownBoard EPD
9. MetsäBoard Prime FBB EB paper profile
10. PaperWise LCA study
11. PEF database, EU average
12. Idemat 2022 database

Glossary

Autoclaving A moulding process for thermoset composite materials that involves laying the material out over an open moulding and using heat to set the material inside a pressure chamber.

Blow moulding A forming process for thermoplastics that starts by injection moulding a pre-form that is then inserted into a blow moulding machine, where it's heated up and inflated to achieve the final shape. Plastic bottles are a common application for this process.

Burst strength A way to measure the strength of paper materials, measured by pressing a test instrument into the material until it bursts.

Compression moulding A forming process that uses pressure to form a material inside a two-part mould, sometimes using heat. Variants of this process exist for many different materials, including plastics, paper, glass and ceramics.

Dip moulding A forming process that's common for rubber products like balloons and gloves. A mould is dipped into liquid rubber, which is then pulled off once the material has solidified.

Direct reduced iron (DRI) and hot briquetted iron (HBI) An alternative process for converting iron ore into iron, without the use of a traditional blast furnace. Renewable hydrogen can be used as the source of energy in this process, making it possible to avoid fossil fuels such as coke altogether. HBI is compacted DRI for more efficient handling, shipping and storage.

Dtex A way of measuring the thickness of textile yarns.

Extrusion A forming process that involves pushing a material through a die to form continuous profiles. Compatible materials include aluminium and plastics.

Feedstocks In the context of plastics, feedstocks refer to the basic ingredients that are used in the production of plastic materials, such as naphtha derived from crude oil or natural gas.

Fibre entanglement process A variety of processes for bonding textile fibres to each other. Leather recycling is one example, where leather waste is shredded into fibre form and reconstituted using a fibre entanglement process.

Glass cullet Crushed glass waste that is used in the production of recycled glass products.

Injection moulding One of the most common ways of forming plastics, this involves heating up the material so that it becomes liquid, before injecting it into a mould where it solidifies in its final shape.

Mechanical recycling Material recycling that involves taking waste apart mechanically through grinding, shredding, crushing or other processes for use in recycled materials.

Overmoulding Injection moulding with more than one material in the same part.

Post-consumer recycled (PCR) Recycled materials that are made with waste that has been collected from consumers, demolished buildings or other sources at the end of its useful life.

Post-industrial recycled (PIR) Recycled materials that are made with factory waste such as offcuts, defect parts and other by-products of industrial processes.

Registration, Evaluation, Authorisation and Restriction of Chemicals (REACH) A set of regulations around the use of harmful chemicals in the European Union. Any materials, additives and secondary finishes that are used in products sold in the EU must be REACH compliant.

Reaction injection moulding A process for injection moulding thermoset plastics that does not rely on heating up the material. Typically a plastic resin and a catalyst are injected into the mould at the same time, starting a chemical reaction that causes the material to cure, or solidify.

Resin transfer moulding A forming process for thermoset composites, where a reinforcement material such as a carbon fibre or glass fibre textile is inserted into a two-part mould, which is then closed before a liquid thermoset resin is injected and cured inside the mould.

Safe service temperature Temperature can have a major impact on the performance of materials – many materials become brittle at low temperatures, and soft and malleable at high temperatures, for example. Safe service temperature refers to the span between the lowest and highest temperatures that a material can be used at without loss of properties.

Sintering A process for bonding powder into solid material using heat or a combination of heat and pressure.

Substances of Very High Concern (SVHC) A list of chemicals and other substances that are known to have serious impacts on human and animal health, as well as the environment in general. Companies and institutions in Europe that use substances that are on the SVHC list must acquire authorization from the European Chemicals Agency (ECHA).

Thermoforming A common sheet forming process that involves heating up the material and then draping it over an open mould, sometimes with the help of a vacuum. Compatible materials include plastics, glass (aka slumping) and aluminium (aka super-plastic forming).

Thermoplastics Plastic materials that become soft and formable when they are heated up, then solidify when they cool down again. This process can be repeated.

Thermosetting plastic / thermoset Plastic materials, often in the form of liquid resin, that are permanently hardened once they have been 'set' using heat, UV light or another catalyst that solidifies the material.

Vulcanization A process for making soft rubber materials harder and more durable.

Resources

Environmental data and LCA tools

There are several free resources that offer environmental data about a wide range of materials. IDEMAT, which stands for Industrial Design & Engineering MATerials database, is a selection of LCI data collected from many different sources and compiled by Sustainable Impact Metrics, a non-profit spin-off from the Delft University of Technology. ÖKOBAUDAT is a free database with environmental data for building materials and processes, available from the German Federal Ministry for Housing, Urban Development and Building. Base IMPACTS®, by the French Agency for Ecological Transition, is another example of a free database that covers many materials.

There are many more commercial databases, but the cost of accessing them can be very high. The Global LCA Data Access network (GLAD) is a search engine that can be used to find out which databases, free and paid for, have environmental data about which materials.

OpenLCA is an open-source software for making LCA studies that's available for free. There are several compatible databases, some of which are free, including the Product Environmental Footprint (PEF) database, which is part of the European Commission's Single Market for Green Products initiative.

For individual material environmental product declarations (EPDs), ECO Platform's ECO Portal is a database with more than 4,000 EPDs from various sources.

Plastic resources

Plastics Europe's Eco-profiles provide environmental data about a very large number of plastic materials and the ingredients that go into making them. The website is worth a visit just for the interactive chart that maps out the entire manufacturing process of each featured material.

Plastic Recyclers Europe, a trade organization, regularly publishes reports with detailed information about recycling in the EU, while European Bioplastics is a good source of reports about renewable plastics.

Design for Recycling, Design from Recycling: Practical Guidelines for Designers is an overview of approaches for circular design with plastics by PolyCE, an initiative that was launched by the European Commission in order to promote plastics recycling.

Textile resources

Textile Exchange is a non-profit organization that has launched several certifications for responsible textile production, as well as several reports, including the annual *Preferred Fiber and Materials Market Report*, which gives an overview of the global textiles market and a wide range of sustainability initiatives and materials.

OEKO-TEX® is a Swiss association behind a number of sustainable textile and leather certifications, as well as the OEKO-TEX® Buying Guide, a database to help designers and brands source responsible textile materials.

Metal resources

World Steel Association and WorldStainless are steel industry organizations that offer environmental data about various steel materials, as well as numerous reports about the steel industry in general.

The International Aluminium Institute offers reports about the environmental impact of aluminium, as well as giving an overview of aluminium recycling and other sustainability topics.

Ceramic and glass resources

Cerame-Unie, the European Ceramic Industry Association, publishes the annual *Inventory of Research and Innovation Projects in the Ceramic Industry*, providing an overview of projects in the EU that aim to reduce the environmental impact of ceramics through more efficient processing, recycling and other initiatives.

The European Container Glass Association (FEVE) regularly publishes reports about the EU packaging glass industry, as well as providing environmental data. Glass for Europe, the trade organization for European flat glass suppliers, offers similar reports and data for the flat glass industry.

Wood resources

The Forest Stewardship Council (FSC) has a publicly available database of its members, which can be searched for suppliers of timber, as well as other forest products like cork and rubber.

The International Union for Conservation of Nature's (IUCN) Red List includes endangered wood species that should not be harvested for timber production.

Paper resources

Beyond timber, the FSC also offers paper certification. FSC-certified paper suppliers can be found in the same database that covers timber suppliers and other forest products. Canopy, a Canadian non-profit, has an eco-paper database with a large number of paper suppliers that use responsibly sourced raw materials and recycled paper in their production.

The Confederation of European Paper Industries (CEPI) offers reports on paper production and recycling in Europe, as well as practical guidelines for circular design with paper and other topics.

Other resources

The Bureau of International Recycling (BIR) publishes an annual report that gives an overview of the global recycling industries, as well as numerous reports that go into detail about the recycling of specific materials, such as plastics, metals and paper.

Index

Page numbers in *italics* refer to illustration captions.

A

AARK Prism Tortoise Watch *60*
Abet Laminati *180*
acrylonitrile butadiene styrene (ABS) 34–5
agriculture 93
  land use 47, 49, 93, 100
Aireal 216, 219–20, *221*
alternative-fibre paper 185
  alternative-fibre packaging paper 206–7
  alternative-fibre paper production 199
  alternative-fibre paperboard 212–13
aluminium 114, 115, 117–18
  aluminium alloys 120, 121
  Hydro REDUXA low-carbon aluminium production 133
  low-carbon aluminium 134–5
  recycled aluminium for casting 126–7
  recycled aluminium for extrusion 122–3
  recycled aluminium for sheet applications 124–5
Alusid *148*
animal-based textiles 72, 73, 92
Arbesser, Arthur *180*
Arctic Blue Gin packaging *208*
autoclaving 18

B

bagasse paper 199, *212*
bamboo paper 199
barrier paper 185, 202–3
  barrier paperboard 185, 210–11
Bentzen, Thomas *165*
Bergans Slingsby Ultra Jacket *98*
Bergström, Lena *136*
biaxially oriented polyethylene (BOPE) 230, *231*
BioBasedTiles® 224, *225*
biodegradability 13–14
  renewable plastics 48, 49
BioFabbrica Bio-Tex™ *106*
Biossance Squalane Oil packaging *212*
Blå Station Röhsska Chair *174*
Blink seating *117, 118*
blow moulding 18
blown glass 156–7
Bolumburu, Pilar 51–2
Bonzano OSB Color samples *177*
borosilicate glass 160–1
Braskem 46
burst strength 192
Butter Chair *30*
By Humankind bag *192*

C

carbon dioxide 10, 220
  CO_2 emissions 16, 46, 96, 114, 115, 132, 162, 163, 224, 226
Carbstone 222, 223
Cavaciocchi, Riccardo 187–8
cellulose 185, 198
  cellulose acetate (CA) 60–1
  cellulose-based synthetic textiles 84–5, 92, 102–3
ceramics 138
  ceramic waste and recycling 145
  ceramic waste streams 145
  origins of ceramic materials 139
  recycling 144–5, 146–51
  sintered stone 150–1
  StoneCycling process 145
  terrazzo 148–9
  waste-based brick clay 146–7
Charlotte McCurdy lab 95–6
  carbon-negative raincoat 96
chemical recycling 74, 217, 226–7
  key types of chemical recycling 227
circularity 12–14
  circular plastics 23
clay 139
  waste-based brick clay 146–7
CleanCloud™ CloudPrime running shoe 222, 223
Coffee Collective bag *231*
composting of plastics 49
compression moulding 42
cork composites 182–3
Cosentino Talyd Vase *150*
cotton textiles 86–7, 108–9
Covestro 36, *40*
Cristal Renew 226
CURV™ 230, *231*
Cyclon Cloudneo running shoe 22, *23*

D

Dekton *150*
DesignByThem *30*
Desmopan® CQ PIR TPU *40*
dip moulding 44
direct reduced iron (DRI) 132

E

Eastman 226, *227*
Ecodear® *98*
ECOLOGO® 184
EcoRub TPRR® 228, *229*
Edelrid SALATHE LITE climbing helmet *231*
Electrolux Pure D9 Green vacuum cleaner *34*
emerging technologies 216
  carbon capture and utilization (CCU) 216, 217, 222–3
  mixed-plastics mechanical recycling 217, 228–9
  mono-material plastic composites 217, 230–1
  natural growth processes 217, 224–5
  plastic chemical recycling 217, 226–7
Emmaljunga NXT90 ERGO stroller *134*
energy 10
Engesvik, Andreas *88*
engineered wood 172–3
  cork composites 182–3
  engineered wood and formaldehyde 173
  high-pressure laminates (HPLs) 180–1
  NAF medium-density fibreboard (MDF) 176–7
  NAF oriented strand board (OSB) 176–7
  NAF plywood 174–5
Envirofoil® *208*
Established & Sons Aura Light *66*
EU Ecolabel 168, 184, 198
European Ceramic Industry Association (Cerame-Unie) 138
European Chemical Industry Council (CEFIC) 226
European Environment Agency 144
European Green Deal 15, 132
EVA 222, 223
EVBox Livo home charging station *62*
Everlane New Day Market Tote *106*
extrusion 18

F

Fairbrics 222, *223*
Fairphone 4 *36, 40*
Fairphone charging cable *82*
feedstocks 46
Fibers Unsorted 77, 77–8, *78*
fibre entanglement process 90
Filosa, Louis *58*

Finisterre Nieuwland 2e Yulex Long Sleeve Swimsuit *70*
Fiskars ReNew Scissors *130*
Fjällräven Tree-Kånken bag *102*
flat glass 156–7
Foersom, Johannes *126*
food crops 47, 93
forest industry 47, 93
  see also wood
Forest Stewardship Council (FSC) 46, 92, 162, 163, 168, 184, 190, 198, 199
Forét Nalgene bottle *227*
formaldehyde 172, 173, 176–7
Friedland, Efrat 25–6
Front *122*

G
Gantri Cantilever Floor Light *58*
Gibson, Sarah *30*
glass 138, 139
  glass cullet 156, 158
  glass waste streams 153
  recycled borosilicate glass 160–1
  recycled flat and blown glass 156–7
  recycled fused glass 158–9
  recycled soda-lime packaging glass 154–5
  recycling 152–3, 154–61
  terrazzo 148–9
Global Organic Textile Standard (GOTS) 72
global warming potential (GWP) 10, 114
  commodity plastics vs technical plastics 19
Google Nest Mini speaker *80*
Granorte *182*
Green Clean toothbrush packaging *196*
GREENGUARD Gold 173
greenhouse gases (GHG) 10, 132, 220
  greenhouse gases absorbed by global forests 163
Grönska barrier paperboard packaging *210*

H
Hand&Eye Studio Io Wall Light *148*
HAY
  Arbour Club Armchair *88*
  Moroccan Vase *156*
Hazen *208*
Helly Hansen 22
high-density polyethylene (HDPE) 30, 56
high-pressure laminates (HPLs) 180–1
Hiort-Lorenzen, Peter *126*
hot briquetted iron (HBI) 132
Hutton, Richard 117–18
HYBRIT fossil-free steel 132, 133, *136*
Hydro REDUXA low-carbon aluminium 132, 133, *134*

I
IKEA
  Kuggis storage box *32*
  SYMFONISK speaker cover *108*
INÉ wallet and powerbank *90*
Initiative for Responsible Mining Assurance (IRMA) 114, 138
injection moulding 18
International Union for Conservation of Nature (IUCN) 168

J
Jordan *196*

K
Karlovasitis, Nick *30*
Kvadrat *88*

L
Lammhults Atlas chair *126*
LanzaTech 222, *223*
latex see rubber
Layer Design Surface Straws *160*
leather 90–1, 106–7, 112–13
Lensvelt *117*
Lenzing AG 74, *84*
Levy, Arik *28*
life-cycle analysis (LCA) 8–9
liquid epoxy resin (LER) 66–7
Loeze soap bar packaging *206*
Lootah, Aljoud *150*
Lorensen, Henrik Taudorf 165–6
low-carbon metals 132–3
  low-carbon aluminium 134–5
  low-carbon steel 136–7
low-density polyethylene (LDPE) 30, 56, 226
lyocell 84, 102

M
Magna Glaskeramik® *158*
Makrolon® PCR *36*
Marcelis, Sabine *66*
Massa, Ward 141–2, 145
materials 6–7, 8, 14–15
  aesthetics 14
  energy, carbon dioxide and materials manufacturing 10
  life-cycle analysis (LCA) 8–9
  longevity 14
  raw materials 10–11
  recycling 12–14
  toxicity 11–12
Materialscout 25–6
Materiom 51–2
  materials from Materiom *53*
McCurdy, Charlotte 95–6
mechanical recycling 20
  mixed-plastics mechanical recycling 217, 228–9
medium-density fibreboard (MDF) 176–7
metals 114
  benchmarking the environmental impact of metals 115
  common aluminium alloys 121
  global annual demand and emissions from metal production 115
  global metal recycling rates 121
  metal waste streams 121
  recycled aluminium for casting 126–7
  recycled aluminium for extrusion 122–3
  recycled aluminium for sheet applications 124–5
  recycled stainless steel 130–1
  recycled steel 128–9
  recycling 120–1
Microsoft Ocean Plastic mouse 228, *229*
Milano Design Week *180*
modal 102
Modus Bob Stool *182*
Montana System modular shelving *177*
moulded paper pulp 185, 196–7, 214–15
Mud Well installation *219*
mycelium 216, 217, 224, *225*

N
natural rubber 44–5, 70–1
natural textiles 73, 92
NCP 15, *128*
no added formaldehyde (NAF) 173
  NAF medium-density fibreboard (MDF) 176–7
  NAF oriented strand board (OSB) 176–7
non-food crops 47, *64*, 93, *100*

non-renewable materials 10–11, 16
  ceramics and glass 138–9
  textiles 73
Norm 1L11-01 trainer *44*
Novelis *124*
nutrient-based materials 47
nylon *see* polyamide (PA)

O

OECD *Global Plastic Outlook* 16
OEKO-TEX® 72, 92
Omni Polymers 228
On Running 22, 23, 222, 223
Orbix 222, 223
oriented strand board (OSB) 176–7
overmoulding 20

P

packaging glass 154–5
paper 184
  alternative-fibre packaging paper 206–7
  alternative-fibre paper production 199
  alternative-fibre paperboard 212–13
  barrier paper 202–3
  barrier paperboard 185, 210–11
  global paper recycling rates 191
  global virgin paper production 199
  moulded paper pulp 214–15
  packaging paper 200–1
  paper certifications 199
  paper materials 185
  paper waste streams 191
  paperboard 208–9
  recycled moulded paper pulp 196–7
  recycled packaging paper 192–3
  recycled paperboard 194–5
  recycling 190–1
  translucent paper 204–5
  uses for recycled paper 191
  virgin paper 198–9
Paper Factor *187*, 187–8, *188*
PaperFoam *214*
PaperWise *206*
Paulsen, Fredrik *174*
PCR (post-consumer recycled waste) 12, 22, 120, *146*
Pearson Lloyd *166*
Philips Eco Conscious Edition Kettle *54*
PIR (post-industrial recycled waste) 12, 22, 120, *146*, *158*, *182*
plant-based raw materials 46–9
  family tree 47
plant-based textiles 72, 73, 92
Plastic Energy 226, 227
plastics 16–18
  building blocks of plastics 18
  cellulose acetate (CA) 60–1
  chemical recycling 217, 226–7
  circular plastics 23
  commodity plastics vs technical plastics 19
  EU plastic recycling capacity 21
  EU plastic recycling rates 21
  family tree 17
  mixed-plastics mechanical recycling 217, 228–9
  mono-material plastic composites 217, 230–1
  natural rubber 70–1
  polylactide (PLA) 58–9
  recycled acrylonitrile butadiene styrene (ABS) 34–5
  recycled natural rubber 44–5
  recycled polyamide (PA) 38–9
  recycled polycarbonate (PC) 36–7
  recycled polyethylene (PE) 30–1
  recycled polyethylene terephthalate (PET) 32–3
  recycled polypropylene (PP) 28–9
  recycled silicone rubber 42–3
  recycled thermoplastic elastomers (TPEs) 40–1
  recycling 20–3, 217, 226–9
  renewable liquid epoxy resin (LER) 66–7
  renewable plastics 46–9
  renewable polyamide (PA) 64–5
  renewable polycarbonate (PC) 62–3
  renewable polyethylene (PE) 56–7
  renewable polypropylene (PP) 54–5
  renewable thermoplastic elastomers (TPEs) 68–9
  specific plastics recycling rates 21
Pleasant Funky Flavour upcycled cap *86*
plywood 174–5
polyamide (PA) 38–9, 64–5, 74, 82–3, 92, 100–1, 230
polycarbonate (PC) 36–7, 62–3
PolyCE *Design for Recycling, Design from Recycling: Practical Guidelines for Designers* 22, 23
polyester (PET) textiles 74, 80–1, 92, 98–9, 104, 226
polyethylene (PE) 28, 30–1, 56–7, 59
polyethylene terephthalate (PET) 22, 28, 30, 32–3, 74, 228, 230
polylactide (PLA) 46–8, 58–9, 92, 104–5
  tea bag *104*
polypropylene (PP) 28–9, 30, 54–5
polystyrene 58
Positive Plastics 25–6
product design 6–7, 8, 14–15
  aesthetics 14
  life-cycle analysis (LCA) 8–9
  longevity 14
  return to circularity 12–14
Programme for the Endorsement of Forest Certification (PEFC) 46, 92, 162, 163, 168, 184, 190, 198, 199
Pyrex *see* borosilicate glass

R

Re-wool *88*
REACH (Registration, Evaluation, Authorisation and Restriction of Chemicals) 12, 18
reaction injection moulding 66
Recovery 152
recycling 12–14, 20–2
  ceramics 144–5
  glass 152–3
  global material recycling rates *13*
  metals 120–1, 122–31
  paper 190–1, 192–7
  plastics 20–3, 28–45, 226–9
  textiles 74–5, 80–91
renewable materials 10–11, 16
  arable land and renewable plastics 49
  global renewable plastics production 49
  metal production 132–3, 134–7
  plastics 46–9, 54–71
  textiles 73, 92–3, 98–113
resin transfer moulding 66
rigid thermoplastics 17
rigid thermosets 17
Ritter Sport Mini packaging *200*
rubber 44–5, 70–1
  silicone rubber 42–3
Rybakken, Daniel *88*

S
S-1500 Chair *15, 128*
SABIC *62*, 228, 229
XENOY™ *229*
safe service temperature 28
Samsung 9000-series LED TV *124*
Sandberg, Saga Maria *210*
Schiphol Airport, Amsterdam 117–18, *117, 118*
Schneider Link-It pen *68*
SCOBY 224, *225*
SCS Global Services Recycled Content Certification 190
Seaman Paper Vela translucent paper bag *204*
sediment 139
Sheldrake, Merlin *Entangled Life* 224
SILICASTONE™ *148*
silicone rubber 42–3
sintering 144
sintered stone 150–1
Snøhetta *15, 128*
soda-lime packaging glass 154–5
Sodeau, Michael *182*
solid wood 168–70, *171*
forest rotation rates 169
solid wood sawing patterns and grain direction 169
Sørensen Leather *112*
Space Copenhagen *112*
Speedo Aquapulse Pro swimming goggles *42*
SSAB 132
candle holder *136*
stainless steel 115, 120, 121
recycled stainless steel 130–1
steel 114, 115, 120, 121
HYBRIT fossil-free steel production 133
low-carbon steel 136–7
recycled steel 128–9
stone 139
sintered stone 150–1
StoneCycling *141*, 141–2, *142*, 145, *146*, *225*
straw 198, *206*, 212
Studio Lionne van Deursen 224, *225*
sustainability 6–7, 8, 14–15
aesthetics of sustainability 14
designing for longevity 14
energy, carbon dioxide and materials manufacturing 10
materials and toxicity 11–12
measuring environmental impact 8–9, 115
origin of raw materials 10–11
return to circularity 12–14
SVHC (Substances of Very High Concern) 12
Swatch® REDVREMYA watch *64*
Swole Panda T-shirt *84*
synthetic textiles 72, 73, 84–5, 102–3

T
TACOIL™ 226, *227*
TAKT 165–6
Cross Chair *166*
Soft Chair *165*
TENCEL™ Lyocell x REFIBRA™ *84*
TePe GOOD™ toothbrush *56*
terrazzo 148–9
Textile Exchange *Preferred Fiber and Materials Market Report* 72, 92
textiles 72
cellulose-based synthetic textiles 102–3
cotton textiles 108–9
fibres, filaments and materials 73
global textile recycling rates 75
leather 112–13
polylactide (PLA) textiles 104–5
recycled cellulose-based synthetic textiles 84–5
recycled cotton textiles 86–7
recycled leather 90–1
recycled polyamide (PA) textiles 82–3
recycled polyester (PET) textiles 80–1
recycled wool textiles 88–9
recycling 74–5
renewable polyamide (PA) textiles 100–1
renewable polyester (PET) textiles 98–9
renewable synthetic leather 106–7
textile waste streams 75
wool textiles 110–11
thermoforming 28
thermoplastic elastomers (TPEs) 17, 40–1, 68–9
thermoplastics 17, 18
thermoset elastomers 17
thermosets 17, 18, 66
timber *see* wood
Tony's Chocolonely wrappers *202*
Total Corbion 46–8
toxicity 11–12
translucent paper 185, 204–5
Tritan Renew 226, 227
TRUCIRCLE™ LEXAN™ renewable polycarbonate *62*

U
UL Solutions 184
UL GREENGUARD 172, 173
ultra-low emitting formaldehyde (ULEF) 173
Unisk *108*
UPM 46
Urban Ears Sumpan earphones *194*

V
Van Dongen, Teresa 216, *219*, 219–20, 222
Vaude Tekoa Biobased Pants *100*
Vestre FOLK bicycle stand *122*
Vetropack Echovai bottle *154*
VIFA OSLO speaker *110*
virgin paper 198–9, 200–15
viscose 84, 102
VITO 222, 223
Vitra Toolbox RE *28*
Vogt-Plastic *28*

W
waste-based materials 47
waste recovery 12–14
WasteBasedBrick® *142*, 145, *146*
Well Kept packaging 224, *225*
Wolf, Michael 77–8
wood 162
engineered wood 172–3, 174–83
global certified forest area 163
global timber trade 163
global wood waste volumes 173
greenhouse gases absorbed by global forests 163
solid wood 168–71
wood-based paper 199
wood waste streams 173
wool textiles 88–9, 110–11

Y
Yulex *70*

Z
Zettle Ocean Reader *38*

Acknowledgements

First and foremost I would like to thank my wife Anne-Charlotte – thank you for cheering me on and for your patience during the countless hours I spent working on this book. Thanks to Liz, Angela and Jane for all your hard work and for making the process of making this book feel less solitary and more like a collaboration. Thanks to Gianni for the great pictures – you are the best Sicilian photographer in Surrey. A big thank you to all the designers, brands and material suppliers that generously gave up their time to give suggestions for improving my text, as well as giving me permission to use images and other resources. In particular, I would like to thank everyone I interviewed for this book – doing the interviews was a highlight and I'm sure readers will find your insights every bit as fascinating and valuable as I do. Last, but not least, many thanks to Chris for the great foreword and for encouraging me to write this book in the first place.

Photo credits

AARK Collective Pty Ltd: 61
Abbie Hughes (Finisterre): 71
Alex Hamstra: 218
Aljoud Lootah Design /Cosentino Group: 151
Aljoud Lootah Design: 187
Arctic Blue Beverages (product owner), Eeva Mäkinen (photo), Luova Työmaa (graphic design), Metsä Board (design for production): 209
Arthur Arbesser for Abet Laminati: 181
Bergans: 99
Bjørnar Øvrebø: 15, 129
Brain of Materials, Envisions; Ronald Smits (photography): 76, 77, 79
Charlotte McCurdy: 94, 95, 97
Daniel Liden: 33, 65, 85, 107, 125, 155, 159, 193, 201
Daniel Schvarcz: 24, 25
Dim Balsem: 140, 141, 143
®Edelrid: 231 bottom
Electrolux: 35
Emmaljunga: 135
Established & Sons: 67
© EVBox: 63
© Fairphone: 37, 41, 83
© Fiskars; Risto Vaurus (photography): 131
FORÉT X NALGENE BOTTLE – BLUE 32 oz (www.foretstudio.dk): 227 left
Gantri (Ian Yang, John Thatcher, Craid Aucutt, Louis Filosa): 59
Gianni Diliberto: 43, 55, 57, 81, 91, 103, 105, 111, 179, 195, 203, 207, 213, 215, 229 top & bottom, 231 top
Gibson Karlo (Sarah Gibson & Nicholas Karlovasitis; design), Pete Daly (photography), DesignByThem (design house): 31
Grönska herb pot: shape developed by Grönska, print design by Saga Maria Sandberg: 211
Hans Boddeke: 219, 221, 223 middle
HAY: 89, 157
Jake Curtis: 183
Jan Snarski (@janmatthewsnarski): 225 bottom right
Jhom Bouwens: 45
Jonas Bjerre-Poulsen: 113
LanzaTech image of Shougang Facility using LanzaTech Technology: 223 bottom
Courtesy of LAYER: 161
Lena Bergström (design) for SSAB: 137
Lensvelt office furniture: 116, 117, 119
Maremosso/Edoardo Delille / Giulia Piermartiri: 189
Materiom: 50, 51, 53
Montana Furniture: 177
Nick Rochowski: 149
On Running; © On AG (on-running.com): 23, 223 top
Orkla Home and Personal Care AS: 197
Paper Factor: 186
Pelle Wahlgren: 127
Plastic Energy Ltd: 227 right
Pleasant (Marc Berliner): 87
Positive Plastics: 27
Rasmus Densøe Studio: 164, 165, 167
Röhsska Chair (produced by Blå Station), Fredrik Paulsen (design), Anders Norrsell (photo): 175
Schneider: 69
Seaman Paper: 205
StoneCycling: 147, 225 top
Unfold by Studio van Deursen: 225 bottom left
Unisk cover for the IKEA Symfonisk Bookshelf speaker (uniskdesign.com): 109
Vaude: 101
Vestre: 123
Vitra: 29
Wikimedia Commons (Anonimski CC0): 171
Zettle/Paypal: 39